MTEL 51

Middle School Mathematics / Science

Teacher Certification Exam

By: Sharon Wynne, M.S
Southern Connecticut State University

"And, while there's no reason yet to panic, I think it's only prudent that we make preparations to panic."

XAMonline, INC.
Boston

XAMonline, Inc.
21 Orient Ave.
Melrose, MA 02176
Toll Free 1-800-509-4128
Email: info@xamonline.com
Web www.xamonline.com
Fax: 1-781-662-9268

Library of Congress Cataloging-in-Publication Data

Wynne, Sharon A.
Middle School Mathematics/Science 51: Teacher Certification / Sharon A. Wynne. -2nd ed.
ISBN 978-1-58197-891-9
1. Middle School Mathematics – Science 51. 2. Study Guides. 3. MTEL
4. Teachers' Certification & Licensure. 5. Careers

Disclaimer:
The opinions expressed in this publication are the sole works of XAMonline and were created independently from the National Education Association, Educational Testing Service, or any State Department of Education, National Evaluation Systems or other testing affiliates.

Between the time of publication and printing, state specific standards as well as testing formats and website information may change that is not included in part or in whole within this product. Sample test questions are developed by XAMonline and reflect similar content as on real tests; however, they are not former tests. XAMonline assembles content that aligns with state standards but makes no claims nor guarantees teacher candidates a passing score. Numerical scores are determined by testing companies such as NES or ETS and then are compared with individual state standards. A passing score varies from state to state.

Printed in the United States of America **œ-1**

MTEL: Middle School Mathematics/Science 51
ISBN: 978-1-58197-891-9

TABLE OF CONTENTS

pg.

Great Study and Testing Tips!

What to study in order to prepare for the subject assessments is the focus of this study guide but equally important is *how* you study.

You can increase your chances of truly mastering the information by taking some simple, but effective steps.

Study Tips:

1. Some foods aid the learning process. Foods such as milk, nuts, seeds, rice, and oats help your study efforts by releasing natural memory enhancers called CCKs (*cholecystokinin*) composed of *tryptophan*, *choline*, and *phenylalanine*. All of these chemicals enhance the neurotransmitters associated with memory. Before studying, try a light, protein-rich meal of eggs, turkey, and fish. All of these foods release the memory enhancing chemicals. The better the connections, the more you comprehend.

Likewise, before you take a test, stick to a light snack of energy boosting and relaxing foods. A glass of milk, a piece of fruit, or some peanuts all release various memory-boosting chemicals and help you to relax and focus on the subject at hand.

2. Learn to take great notes. A by-product of our modern culture is that we have grown accustomed to getting our information in short doses (i.e. TV news sound bites or USA Today style newspaper articles.)

Consequently, we've subconsciously trained ourselves to assimilate information better in neat little packages. If your notes are scrawled all over the paper, it fragments the flow of the information. Strive for clarity. Newspapers use a standard format to achieve clarity. Your notes can be much clearer through use of proper formatting. A very effective format is called the *"Cornell Method."*

> Take a sheet of loose-leaf lined notebook paper and draw a line all the way down the paper about 1-2" from the left-hand edge.
>
> Draw another line across the width of the paper about 1-2" up from the bottom. Repeat this process on the reverse side of the page.

Look at the highly effective result. You have ample room for notes, a left hand margin for special emphasis items or inserting supplementary data from the textbook, a large area at the bottom for a brief summary, and a little rectangular space for just about anything you want.

3. **Get the concept then the details.** Too often we focus on the details and don't gather an understanding of the concept. However, if you simply memorize only dates, places, or names, you may well miss the whole point of the subject.

A key way to understand things is to put them in your own words. If you are working from a textbook, automatically summarize each paragraph in your mind. If you are outlining text, don't simply copy the author's words.

Rephrase them in your own words. You remember your own thoughts and words much better than someone else's, and subconsciously tend to associate the important details to the core concepts.

4. **Ask Why?** Pull apart written material paragraph by paragraph and don't forget the captions under the illustrations.

Example: If the heading is "Stream Erosion", flip it around to read "Why do streams erode?" Then answer the questions.

If you train your mind to think in a series of questions and answers, not only will you learn more, but it also helps to lessen the test anxiety because you are used to answering questions.

5. **Read for reinforcement and future needs.** Even if you only have 10 minutes, put your notes or a book in your hand. Your mind is similar to a computer; you have to input data in order to have it processed. *By reading, you are creating the neural connections for future retrieval.* The more times you read something, the more you reinforce the learning of ideas.

Even if you don't fully understand something on the first pass, *your mind stores much of the material for later recall.*

6. **Relax to learn so go into exile.** Our bodies respond to an inner clock called biorhythms. Burning the midnight oil works well for some people, but not everyone.

If possible, set aside a particular place to study that is free of distractions. Shut off the television, cell phone, and pager and exile your friends and family during your study period.

If you really are bothered by silence, try background music. Light classical music at a low volume has been shown to aid in concentration over other types. Music that evokes pleasant emotions without lyrics is highly suggested. Try just about anything by Mozart. It relaxes you.

7. Use arrows not highlighters. At best, it's difficult to read a page full of yellow, pink, blue, and green streaks. Try staring at a neon sign for a while and you'll soon see that the horde of colors obscure the message.

A quick note, a brief dash of color, an underline, and an arrow pointing to a particular passage is much clearer than a horde of highlighted words.

8. Budget your study time. Although you shouldn't ignore any of the material, ***allocate your available study time in the same ratio that topics may appear on the test.***

Testing Tips:

1. **Get smart, play dumb. Don't read anything into the question.** Don't make an assumption that the test writer is looking for something else than what is asked. Stick to the question as written and don't read extra things into it.

2. **Read the question and all the choices *twice* before answering the question.** You may miss something by not carefully reading, and then re-reading both the question and the answers.

If you really don't have a clue as to the right answer, leave it blank on the first time through. Go on to the other questions, as they may provide a clue as to how to answer the skipped questions.

If later on, you still can't answer the skipped ones . . . ***Guess.*** The only penalty for guessing is that you *might* get it wrong. Only one thing is certain; if you don't put anything down, you will get it wrong!

3. **Turn the question into a statement.** Look at the way the questions are worded. The syntax of the question usually provides a clue. Does it seem more familiar as a statement rather than as a question? Does it sound strange?

By turning a question into a statement, you may be able to spot if an answer sounds right, and it may also trigger memories of material you have read.

4. **Look for hidden clues.** It's actually very difficult to compose multiple-foil (choice) questions without giving away part of the answer in the options presented.

In most multiple-choice questions you can often readily eliminate one or two of the potential answers. This leaves you with only two real possibilities and automatically your odds go to Fifty-Fifty for very little work.

5. **Trust your instincts.** For every fact that you have read, you subconsciously retain something of that knowledge. On questions that you aren't really certain about, go with your basic instincts. **Your first impression on how to answer a question is usually correct.**

6. **Mark your answers directly on the test booklet.** Don't bother trying to fill in the optical scan sheet on the first pass through the test.

Just be very careful not to miss-mark your answers when you eventually transcribe them to the scan sheet.

7. **Watch the clock!** You have a set amount of time to answer the questions. Don't get bogged down trying to answer a single question at the expense of 10 questions you can more readily answer.

SUBAREA I. NUMBER SENSE AND OPERATIONS

Competency 0001 Understand the structure of numeration systems and multiple representations of numbers.

Prime numbers are numbers that can only be factored into 1 and the number itself. When factoring into prime factors, all the factors must be numbers that cannot be factored again (without using 1). Initially numbers can be factored into any 2 factors. Check each resulting factor to see if it can be factored again. Continue factoring until all remaining factors are prime. This is the list of prime factors. Regardless of which way the original number was factored, the final list of prime factors will always be the same.

Example: Factor 30 into prime factors.

Divide by 2 as many times as you can, then by 3, then by other successive primes as required.

$2 \cdot 2 \cdot 2 \cdot 2 \cdot 2 \cdot 2 \cdot 2 \cdot 2 \cdot 2 \cdot 2 \cdot 2 \cdot 2 \cdot 2 \cdot 2 \cdot 2$

Factor 30 into any 2 factors.

$5 \cdot 6$	Now factor the 6.
$5 \cdot 2 \cdot 3$	These are all prime factors.

Factor 30 into any 2 factors.

$3 \cdot 10$	Now factor the 10.
$3 \cdot 2 \cdot 5$	These are the same prime factors even though the original factors were different.

Example: Factor 240 into prime factors.

Factor 240 into any 2 factors.

$24 \cdot 10$	Now factor both 24 and 10.
$4 \cdot 6 \cdot 2 \cdot 5$	Now factor both 4 and 6.
$2 \cdot 2 \cdot 2 \cdot 3 \cdot 2 \cdot 5$	These are prime factors.

This can also be written as $2^4 \cdot 3 \cdot 5$

GCF is the abbreviation for the **greatest common factor**. The GCF is the largest number that is a factor of all the numbers given in a problem. The GCF can be no larger than the smallest number given in the problem. If no other number is a common factor, then the GCF will be the number 1. To find the GCF, list all possible factors of the smallest number given (include the number itself). Starting with the largest factor (which is the number itself), determine if it is also a factor of all the other given numbers. If so, that is the GCF. If that factor doesn't work, try the same method on the next smaller factor. Continue until a common factor is found. That is the GCF. Note: There can be other common factors besides the GCF.

<u>Example:</u> Find the GCF of 12, 20, and 36.

The smallest number in the problem is 12. The factors of 12 are 1,2,3,4,6 and 12. 12 is the largest factor, but it does not divide evenly into 20. Neither does 6, but 4 will divide into both 20 and 36 evenly.
Therefore, 4 is the GCF.

<u>Example:</u> Find the GCF of 14 and 15.

Factors of 14 are 1,2,7 and 14. 14 is the largest factor, but it does not divide evenly into 15. Neither does 7 or 2. Therefore, the only factor common to both 14 and 15 is the number 1, the GCF.

LCM is the abbreviation for **least common multiple**. The least common multiple of a group of numbers is the smallest number that all of the given numbers will divide into. The least common multiple will always be the largest of the given numbers or a multiple of the largest number.

<u>Example:</u> Find the LCM of 20, 30 and 40.

The largest number given is 40, but 30 will not divide evenly into 40. The next multiple of 40 is 80 (2 x 40), but 30 will not divide evenly into 80 either. The next multiple of 40 is 120. 120 is divisible by both 20 and 30, so 120 is the LCM (least common multiple).

<u>Example:</u> Find the LCM of 96, 16 and 24.

The largest number is 96. 96 is divisible by both 16 and 24, so 96 is the LCM.

Divisibility Tests

a. A number is divisible by 2 if that number is an even number (which means it ends in 0,2,4,6 or 8).

1,354 ends in 4, so it is divisible by 2. 240,685 end in a 5, so it is not divisible by 2.

b. A number is divisible by 3 if the sum of its digits is evenly divisible by 3.

The sum of the digits of 964 is 9+6+4 = 19. Since 19 is not divisible by 3, neither is 964. The digits of 86,514 is 8+6+5+1+4 = 24. Since 24 is divisible by 3, 86,514 is also divisible by 3.

c. A number is divisible by 4 if the number in its last 2 digits is evenly divisible by 4.

The number 113,336 ends with the number 36 in the last 2 columns. Since 36 is divisible by 4, then 113,336 is also divisible by 4.

The number 135,627 ends with the number 27 in the last 2 columns. Since 27 is not evenly divisible by 4, then 135,627 is also not divisible by 4.

d. A number is divisible by 5 if the number ends in either a 5 or a 0.

225 ends with a 5 so it is divisible by 5. The number 470 is also divisible by 5 because its last digit is a 0. 2,358 is not divisible by 5 because its last digit is an 8, not a 5 or a 0.

e. A number is divisible by 6 if the number is even and the sum of its digits is evenly divisible by 3 or 6.

4,950 is an even number and its digits add to 18. (4+9+5+0 = 18) Since the number is even and the sum of its digits is 18 (which is divisible by 3 and/or 6), then 4950 is divisible by 6. 326 is an even number, but its digits add up to 11. Since 11 is not divisible by 3 or 6, then 326 is not divisible by 6. 698,135 is not an even number, so it cannot possibly be divided evenly by 6.

f. A number is divisible by 8 if the number in its last 3 digits is evenly divisible by 8.

The number 113,336 ends with the 3-digit number 336 in the last 3 columns. Since 336 is divisible by 8, then 113,336 is also divisible by 8. The number 465,627 ends with the number 627 in the last 3 columns. Since 627 is not evenly divisible by 8, then 465,627 is also not divisible by 8.

g. A number is divisible by 9 if the sum of its digits is evenly divisible by 9.

The sum of the digits of 874 is 8+7+4 = 19. Since 19 is not divisible by 9, neither is 874. The digits of 116,514 is 1+1+6+5+1+4 = 18. Since 18 is divisible by 9, 116,514 is also divisible by 9.

h. A number is divisible by 10 if the number ends in the digit 0.

305 ends with a 5 so it is not divisible by 10. The number 2,030,270 is divisible by 10 because its last digit is a 0. 42,978 is not divisible by 10 because its last digit is an 8, not a 0.

i. Why these rules work.

All even numbers are divisible by 2 by definition. A 2-digit number (with T as the tens digit and U as the ones digit) has as its sum of the digits, T + U. Suppose this sum of T + U is divisible by 3. Then it equals 3 times some constant, K. So, T + U = 3K. Solving this for U, U = 3K - T. The original 2 digit number would be represented by 10T + U. Substituting 3K - T in place of U, this 2-digit number becomes 10T + U = 10T + (3K - T) = 9T + 3K. This 2-digit number is clearly divisible by 3, since each term is divisible by 3. Therefore, if the sum of the digits of a number is divisible by 3, then the number itself is also divisible by 3. Since 4 divides evenly into 100, 200, or 300, 4 will divide evenly into any amount of hundreds. The only part of a number that determines if 4 will divide into it evenly is the number in the last 2 columns. Numbers divisible by 5 ends in 5 or 0. This is clear if you look at the answers to the multiplication table for 5. Answers to the multiplication table for 6 are all even numbers. Since 6 factors into 2 times 3, the divisibility rules for 2 and 3 must both work. Any number of thousands is divisible by 8. Only the last 3 columns of the number determine whether or not it is divisible by 8. A 2 digit number (with T as the tens digit and U as the ones digit) has as its sum of the digits, T + U. Suppose this sum of T + U is divisible by 9. Then it equals 9 times some constant, K. So, T + U = 9K. Solving this for U, U = 9K - T.

Composite numbers are whole numbers that have more than 2 different factors. For example 9 is composite because besides factors of 1 and 9, 3 is also a factor. 70 is also composite because besides the factors of 1 and 70, the numbers 2,5,7,10,14, and 35 are also all factors.

Mathematical operations include addition, subtraction, multiplication and division.

Addition can be indicated by the expressions: sum, greater than, and, more than, increased by, added to.

Subtraction can be expressed by: difference, fewer than, minus, less than, decreased by.

Multiplication is shown by: product, times, multiplied by, twice.

Division is used for: quotient, divided by, ratio.

Examples:

7 added to a number	$n + 7$
a number decreased by 8	$n - 8$
12 times a number divided by 7	$12n \div 7$
28 less than a number	$n - 28$
the ratio of a number to 55	$\frac{n}{55}$
4 times the sum of a number and 21	$4(n + 21)$

A **numeration system** is a set of numbers represented a by a set of symbols (numbers, letters, or pictographs). Sets can have different bases of numerals within the set. Instead of our base 10, a system may use any base set from 2 on up. The position of the number in that representation defines its exact value. Thus, the numeral 1 has a value of ten when represented as “10”. Early systems, such as the Babylonian used position in relation to other numerals or column position for this purpose since they lacked a zero to represent an empty position.

Competency 0002 Understand principles and operations related to integers, fractions, decimals, percents, ratios, and proportions.

The Order of Operations are to be followed when evaluating algebraic expressions. Remember the mnemonic PEMDAS (Please Excuse My Dear Aunt Sally) to follow these steps in order:

1. Simplify inside grouping characters such as parentheses, brackets, radicals, fraction bars, etc.

2. Multiply out expressions with exponents.

3. Do multiplication or division from left to right.

4. Do addition or subtraction from left to right.

Samples of simplifying expressions with exponents:

$(-2)^3 = -8$ $\qquad -2^3 = -8$

$(-2)^4 = 16$ $\qquad -2^4 = -16$ $\qquad$ Note change of sign.

$(2/3)^3 = 8/27$

$5^0 = 1$

$4^{-1} = 1/4$

Inverse operations are operations that "undo" each other. Addition and subtraction are inverse operations since 3 + 8 = 11 and 11 - 8 = 3. Similarly, multiplication and division are inverse operations since 2 × 7 = 14 and 14 ÷ 2 = 7.

Inverse operations are used to solve equations.

Example 1: $x + 12 = 17$ Subtract 12 from each side of the equation since the inverse of addition is subtraction.

$- 12 \quad -12$

$x = 5$

Example 2: $4x = {}^{-}20$ Divide both sides of the equation by 4 since the inverse of multiplication is division.

$$\frac{4x}{4} = \frac{-20}{4}$$

$x = {}^{-}5$

The **absolute value** of a real number is the positive value of that number.

$|x|$ = x when $x \geq 0$ and
$|x|$ = x when $x < 0$.

<u>Examples</u>: $|7| = 7$ $\qquad |-13| = 13$

Examining the change in **area or volume of a given figure** requires first to find the existing area given the original dimensions and then finding the new area given the increased dimensions.

Sample problem:

Given the rectangle below determine the change in area if the length is increased by 5 and the width is increased by 7.

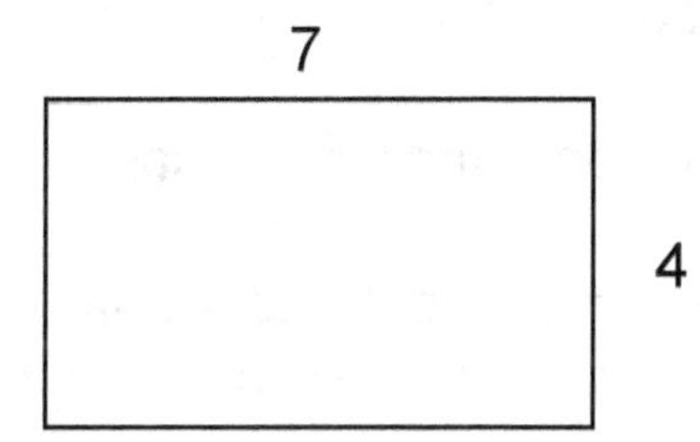

Draw and label a sketch of the new rectangle.

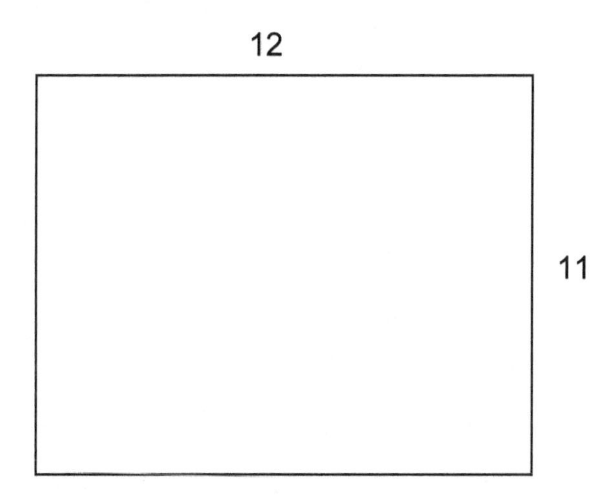

Find the areas.

Area of original = LW
= (7)(4)
= 28 units^2

Area of enlarged shape = LW
= (12)(11)
= 132 units^2

The change in area is 132 – 28 = 104 units^2.

Computers use a base of 2 but combine it into 4 units called a byte to function in base 16 (hexadecimal). A base of 8 (octal) was also used by older computers.

A base of 2 uses only 0 and 1.

Decimal Binary Conversion		
Decimal	Binary	Place Value
1	1	2^0
2	10	2^1
4	100	2^2
8	1000	2^3

Thus, 9 in Base 10 becomes 1001 in Base 2.

9+4 = 13 (Base 10) becomes 1001 + 100 = 1101 (Base 2).

Fractions, ratios and other functions alter in the same way.

Computers use a base of 2 but combine it into 4 units called a byte to function in base 16 (hexadecimal). A base of 8 (octal) was also used by older computers.

Competency 0003 Understand and solve problems involving integers, fractions, decimals, percents, ratios, and proportions.

The unit rate for purchasing an item is its price divided by the number of pounds/ounces, etc. in the item. The item with the lower unit rate is the lower price.

Example: Find the item with the best unit price:

$1.79 for 10 ounces
$1.89 for 12 ounces
$5.49 for 32 ounces

$$\frac{1.79}{10} = 0.179 \text{ per ounce} \quad \frac{1.89}{12} = 0.1575 \text{ per ounce} \quad \frac{5.49}{32} = 0.172 \text{ per ounce}$$

$1.89 for 12 ounces is the best price.

A second way to find the better buy is to make a proportion with the price over the number of ounces, etc. Cross multiply the proportion, writing the products above the numerator that is used. The better price will have the smaller product.

Example: Find the better buy:

$8.19 for 40 pounds or $4.89 for 22 pounds

Find the unit price.

$$\frac{40}{8.19} = \frac{1}{x} \qquad \frac{22}{4.89} = \frac{1}{x}$$

$$40x = 8.19 \qquad 22x = 4.89$$

$$x = 0.20475 \qquad x = 0.22\overline{27}$$

Since $0.20475 < 0.22\overline{27}$, $8.19 is less and is a better buy.

To find the amount of sales tax on an item, change the percent of sales tax into an equivalent decimal number. Then multiply the decimal number times the price of the object to find the sales tax. The total cost of an item will be the price of the item plus the sales tax.

Example: A guitar costs $120 plus 7% sales tax. How much are the sales tax and the total bill?

7% = .07 as a decimal (.07)(120) = $8.40 sales tax
$120 + $8.40 = $128.40 ← total cost

An alternative method to find the total cost is to multiply the price times the factor 1.07 (price + sales tax):

$$\$120 \times 1.07 = \$8.40$$

This gives you the total cost in fewer steps.

Example: A suit costs $450 plus 6½% sales tax. How much are the sales tax and the total bill?

6½% = .065 as a decimal
(.065)(450) = $29.25 sales tax
$450 + $29.25 = $479.25 ← total cost

An alternative method to find the total cost is to multiply the price times the factor 1.065 (price + sales tax):

$$\$450 \times 1.065 = \$479.25$$

This gives you the total cost in fewer steps.

A **ratio** is a comparison of 2 numbers. If a class had 11 boys and 14 girls, the ratio of boys to girls could be written one of 3 ways:

$$11:14 \quad \text{or} \quad 11 \text{ to } 14 \quad \text{or} \quad \frac{11}{14}$$

The ratio of girls to boys is:

$$14:11,\ 14 \text{ to } 11 \text{ or } \frac{14}{11}$$

Ratios can be reduced when possible. A ratio of 12 cats to 18 dogs would reduce to 2:3, 2 to 3 or $2/3$.

Note: Read ratio questions carefully. Given a group of 6 adults and 5 children, the ratio of children to the entire group would be 5:11.
A **proportion** is an equation in which a fraction is set equal to another. To solve the proportion, multiply each numerator times the other fraction's denominator. Set these two products equal to each other and solve the resulting equation. This is called **cross-multiplying** the proportion.

Example: $\frac{4}{15}=\frac{x}{60}$ is a proportion.

To solve this, cross multiply.

$$(4)(60)=(15)(x)$$
$$240=15x$$
$$16=x$$

Example: $\frac{x+3}{3x+4}=\frac{2}{5}$ is a proportion.

To solve, cross multiply.

$$5(x+3)=2(3x+4)$$
$$5x+15=6x+8$$
$$7=x$$

Example: $\frac{x+2}{8}=\frac{2}{x-4}$ is another proportion.

To solve, cross multiply.

$$(x+2)(x-4)=8(2)$$
$$x^2-2x-8=16$$
$$x^2-2x-24=0$$
$$(x-6)(x+4)=0$$
$$x=6 \text{ or } x={}^{-}4$$

Both answers work.

Fractions, decimals, and percents can be used interchangeably within problems.

→To change a percent into a decimal, move the decimal point two places to the left and drop off the percent sign.

→To change a decimal into a percent, move the decimal two places to the right and add on a percent sign.

→To change a fraction into a decimal, divide the numerator by the denominator.

→To change a decimal number into an equivalent fraction, write the decimal part of the number as the fraction's numerator. As the fraction's denominator use the place value of the last column of the decimal. Reduce the resulting fraction as far as possible.

Example: J.C. Nickels has Hunch jeans $1/4$ off the usual price of $36.00. Shears and Roadkill have the same jeans 30% off their regular price of $40. Find the cheaper price.

$1/4$ = .25 so .25(36) = $9.00 off $36 - 9 = $27 sale price
30% = .30 so .30(40) = $12 off $40 - 12 = $28 sale price

The price at J.C Nickels is actually lower.

Elapsed time problems are usually one of two types. One type of problem is the elapsed time between 2 times given in hours, minutes, and seconds. The other common type of problem is between 2 times given in months and years.

For any time of day past noon, change it into military time by adding 12 hours. For instance, 1:15 p.m. would be 13:15. Remember when you borrow a minute or an hour in a subtraction problem that you have borrowed 60 more seconds or minutes.

Example: Find the time from 11:34:22 a.m. until 3:28:40 p.m.

First change 3:28:40 p.m. to 15:28:40 p.m.
Now subtract - 11:34:22 a.m.
:18

Borrow an hour and add 60 more minutes. Subtract
14:88:40 p.m.
- 11:34:22 a.m.
3:54:18 ↔ 3 hours, 54 minutes, 18 seconds

Example: John lived in Arizona from September 91 until March 95. How long is that?

		year	month
March 95	=	95	03
September 91	= -	91	09

Borrow a year, change it into 12 more months, and subtract.

		year	month
March 95	=	94	15
September 91	= -	91	09
		3 yr	6 months

Example: A race took the winner 1 hr. 58 min. 12 sec. on the first half of the race and 2 hr. 9 min. 57 sec. on the second half of the race. How much time did the entire race take?

```
   1 hr. 58 min. 12 sec.
 + 2 hr.  9 min. 57 sec.     Add these numbers
   3 hr. 67 min. 69 sec.
 + 1 min -60 sec.            Change 60 seconds to 1 min.
   3 hr. 68 min.  9 sec.
 + 1 hr.-60 min.        .    Change 60 minutes to 1 hr.
   4 hr.  8 min.  9 sec.  ←final answer
```

Estimation and approximation may be used to check the reasonableness of answers.

Example: Estimate the answer.

$$\frac{58 \times 810}{1989}$$

58 becomes 60, 810 becomes 800 and 1989 becomes 2000.

$$\frac{60 \times 800}{2000} = 24$$

Word problems: An estimate may sometimes be all that is needed to solve a problem.

Example: Janet goes into a store to purchase a CD on sale for $13.95. While shopping, she sees two pairs of shoes, prices $19.95 and $14.50. She only has $50. Can she purchase everything? (Assume there is no sales tax.)

Solve by rounding:

```
$19.95→$20.00
$14.50→$15.00
$13.95→$14.00
       $49.00     Yes, she can purchase the CD and the shoes.
```

Competency 0004 Understand the properties of real numbers and the real number system.

The real number properties are best explained in terms of a small set of numbers. For each property, a given set will be provided.

Axioms of Addition

Closure—For all real numbers a and b, a + b is a unique real number.

Associative—For all real numbers a, b, and c, (a + b) + c = a + (b + c).

Additive Identity—There exists a unique real number 0 (zero) such that a + 0 = 0 + a = a for every real number a.

Additive Inverses—For each real number *a*, there exists a real number *–a* (the opposite of *a*) such that $a + (-a) = (-a) + a = 0$.

Commutative—For all real numbers *a* and *b*, $a + b = b + a$.

Axioms of Multiplication

Closure—For all real numbers *a* and *b*, *ab* is a unique real number.

Associative—For all real numbers *a, b,* and *c*, $(ab)c = a(bc)$.

Multiplicative Identity—There exists a unique nonzero real number 1 (one) such that $1 \bullet a = a$, $a \bullet 1 = a$.

Multiplicative Inverses—For each nonzero real number, there exists a real number $1/a$ (the reciprocal of a) such that $a(1/a) = (1/a)a = 1$.

Commutative—For all real numbers *a* and *b*, $ab = ba$.

The Distributive Axiom of Multiplication over Addition

For all real numbers *a, b,* and *c*, $a(b + c) = ab + ac$.

A. **Natural numbers**--the counting numbers, $1, 2, 3, ...$

B. **Whole numbers**--the counting numbers along with zero, $0, 1, 2...$

C. **Integers**--the counting numbers, their opposites, and zero, $..., {}^{-}1, 0, 1, ...$

D. **Rationals**--all of the fractions that can be formed from the whole numbers. Zero cannot be the denominator. In decimal form, these numbers will either be terminating or repeating decimals. Simplify square roots to determine if the number can be written as a fraction.

E. **Irrationals**--real numbers that cannot be written as a fraction. The decimal forms of these numbers are neither terminating nor repeating. Examples: $\pi, e, \sqrt{2}$, etc.

F. **Real numbers**--the set of numbers obtained by combining the rationals and irrationals. Complex numbers, i.e. numbers that involve i or $\sqrt{{}^{-}1}$, are not real numbers.

The **Denseness Property** of real numbers states that, if all real numbers are ordered from least to greatest on a number line, there is an infinite set of real numbers between any two given numbers on the line.

Example:

Between 7.6 and 7.7, there is the rational number 7.65 in the set of real numbers.
Between 3 and 4 there exists no other natural number.

The **exponent form** is a shortcut method to write repeated multiplication. The **base** is the factor. The **exponent** tells how many times that number is multiplied by itself.

The following are basic rules for exponents:
$a^1 = a$ for all values of a; thus $17^1 = 17$

$b^0 = 1$ for all values of b; thus $24^0 = 1$

$10^n = 1$ with n zeros; thus $10^6 = 1,000,000$

To change a number into **scientific notation**, move the decimal point so that only a single digit is to the left of the decimal point. Drop off any trailing zeros. Multiply this number times 10 to a power. The power is the number of positions that the decimal point is moved. The power is negative if the original number is a decimal number between 1 and -1. Otherwise the power is positive.

Example: Change into scientific notation:

4,380,000,000	Move decimal behind the 4
4.38	Drop trailing zeros.
$4.38 \times 10^{?}$	Count positions that the decimal point has moved.
4.38×10^{9}	This is the answer.
$^{-}.0000407$	Move decimal behind the 4
$^{-}4.07$	Count positions that the decimal point has moved.
4.07×10^{-5}	Note negative exponent.

If a number is already in scientific notation, it can be changed back into regular decimal form. If the exponent on the number 10 is negative, move the decimal point to the left that number of places.. If the exponent on the number 10 is positive, move the decimal point to the right.

Example: Change back into decimal form:

3.448×10^{-2}	Move decimal point 2 places left, since exponent is negative.
.03448	This is the answer.
6×10^{4}	Move decimal point 4 places right, since exponent is positive.
60,000	This is the answer.

To add or subtract in scientific notation, the exponents must be the same. Then add the decimal portions, keeping the power of 10 the same. Then move the decimal point and adjust the exponent to keep the number to the left of the decimal point to a single digit.

Example:

6.22×10^3	
$+\ 7.48 \times 10^3$	Add these as is.
13.70×10^3	Now move decimal 1 more place to the left and
1.37×10^4	add 1 more exponent.

To multiply or divide in scientific notation, multiply or divide the decimal part of the numbers. In multiplication, add the exponents of 10. In division, subtract the exponents of 10. Then move the decimal point and adjust the exponent to keep the number to the left of the decimal point to a single digit.

Example:

$(5.2 \times 10^5)(3.5 \times 10^2)$	Multiply $5.2 \cdot 3.5$
18.2×10^7	Add exponent
1.82×10^8	Move decimal point and increase the exponent by 1.

Example:

$\dfrac{(4.1076 \times 10^3)}{2.8 \times 10^{-4}}$	Divide 4.1076 by 2.8
	Subtract $3 - (^-4)$
1.467×10^7	

SUBAREA II. PATTERNS, RELATIONS, AND ALGEBRA

Competency 0005 Understand and use patterns to model and solve problems.

Kepler discovered a relationship between the average distance of a planet from the sun and the time it takes the planet to orbit the sun.

The following table shows the data for the six planets closest to the sun:

	Mercury	Venus	Earth	Mars	Jupiter	Saturn
Average distance, x	0.387	0.723	1	1.523	5.203	9.541
x^3	0.058	.378	1	3.533	140.852	868.524
Time, y	0.241	0.615	1	1.881	11.861	29.457
y^2	0.058	0.378	1	3.538	140.683	867.715

Looking at the data in the table, we can assume that $x^3 \simeq y^2$.

We can conjecture the following function for Kepler's relationship:

$$y = \sqrt{x^3}$$

The **iterative process** involves repeated use of the same steps. A **recursive function** is an example of the iterative process. A recursive function is a function that requires the computation of all previous terms in order to find a subsequent term. Perhaps the most famous recursive function is the **Fibonacci sequence**. This is the sequence of numbers 1,1,2,3,5,8,13,21,34 ... for which the next term is found by adding the previous two terms.

Example: Find the recursive formula for the sequence 1, 3, 9, 27, 81...

We see that any term other than the first term is obtained by multiplying the preceding term by 3. Then, we may express the formula in symbolic notation as

$$a_n = 3a_{n-1},\ a_1 = 1,$$

where a represents a term, the subscript n denotes the place of the term in the sequence and the subscript $n-1$ represents the preceding term.

A **linear function** is a function defined by the equation $f(x) = mx + b$.

Example: A model for the distance traveled by a migrating monarch butterfly looks like $f(t) = 80t$, where t represents time in days. We interpret this to mean that the average speed of the butterfly is 80 miles per day and distance traveled may be computed by substituting the number of days traveled for t. In a linear function, there is a **constant** rate of change.

The standard form of a **quadratic function** is $f(x) = ax^2 + bx + c$.

Example: What patterns appear in a table for $y = x^2 - 5x + 6$?

x	y
0	6
1	2
2	0
3	0
4	2
5	6

We see that the values for y are **symmetrically** arranged.

An **exponential function** is a function defined by the equation $y = ab^x$, where a is the starting value, b is the growth factor, and x tells how many times to multiply by the growth factor.

Example: $y = 100(1.5)^x$

x	y
0	100
1	150
2	225
3	337.5
4	506.25

This is an **exponential** or multiplicative pattern of growth.

The **iterative process** involves repeated use of the same steps. A **recursive function** is an example of the iterative process. A recursive function is a function that requires the computation of all previous terms in order to find a subsequent term. Perhaps the most famous recursive function is the **Fibonacci sequence**. This is the sequence of numbers 1,1,2,3,5,8,13,21,34 … for which the next term is found by adding the previous two terms.

Competency 0006 Understand how to manipulate and simplify algebraic expressions and translate problems into algebraic notation.

An equation may often offer a simple solution to a word problem.

Example: Mark and Mike are twins. Three times Mark's age plus four equals four times Mike's age minus 14. How old are the boys?

Since the boys are twins, their ages are the same. "Translate" the English into Algebra.

Let x = their age
$3x + 4 = 4x - 14$
$18 = x$

The boys are each 18 years old.

There may be times when it is necessary or easier to write two equations in two variables to solve a word problem.

Example: The sum of two numbers is 48. One number is three times the other. Find the numbers.

Let x = smaller number
y = larger number

$x + y = 48$
$y = 3x$

Solve the linear equations by substitution.

$x + 3x = 48$
$4x = 48$
$x = 12$ so $y = 36$

The numbers are 12 and 36.

DIRECT VARIATION

A **direct variation** can be expressed by the formula
$y = kx$ where k is a constant, $k \neq 0$.

Example: If y varies directly as x and $y = -8$ when $x = 4$, find y when $x = 11$

First find the **constant of variation**, k.

$$k = \frac{y}{x} = \frac{-8}{4} = -2$$

The write an equation with k = -2

$y = -2(11)$
$y = -22$

INDIRECT VARIATION

An **indirect variation** can be expressed by the formula
$xy = k$, where k is a constant, $k \neq 0$.

Example: If y varies inversely as x and $y = 20$ when $x = -4$, find y when $x = 14$

$k = 20(-4) = -80$

Similarly, write and equations with k = -80
$y = \frac{k}{x}$
$y = \frac{-80}{14} = -5\frac{5}{7}$

Competency 0007 Understand properties of functions and relations.

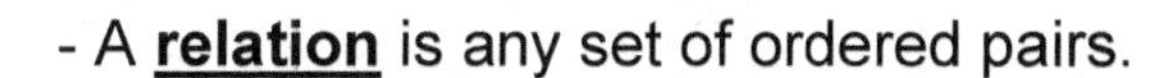

- A **relation** is any set of ordered pairs.

- The **domain** of a relation is the set made of all the first coordinates of the ordered pairs.

- The **range** of a relation is the set made of all the second coordinates of the ordered pairs.

- A **function** is a relation in which different ordered pairs have different first coordinates. (No x values are repeated.)

- A **mapping** is a diagram with arrows drawn from each element of the domain to the corresponding elements of the range. If 2 arrows are drawn from the same element of the domain, then it is not a function.

- On a graph, use the **vertical line test** to look for a function. If any vertical line intersects the graph of a relation in more than one point, then the relation is not a function.

1. Determine the domain and range of this mapping.

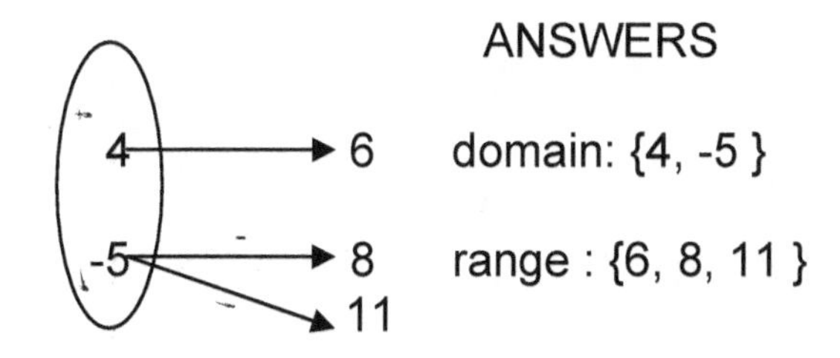

ANSWERS

domain: {4, -5 }

range : {6, 8, 11 }

2. Determine which of these are functions:

a. $\{(1,{}^{-}4),(27,1)(94,5)(2,{}^{-}4)\}$

b. $f(x) = 2x - 3$

c. $A = \{(x,y) \mid xy = 24\}$

d. $y = 3$

e. $x = {}^{-}9$

f. $\{(3,2),(7,7),(0,5),(2,{}^{-}4),(8,{}^{-}6),(1,0),(5,9),(6,{}^{-}4)\}$

3. Determine the domain and range of this graph.

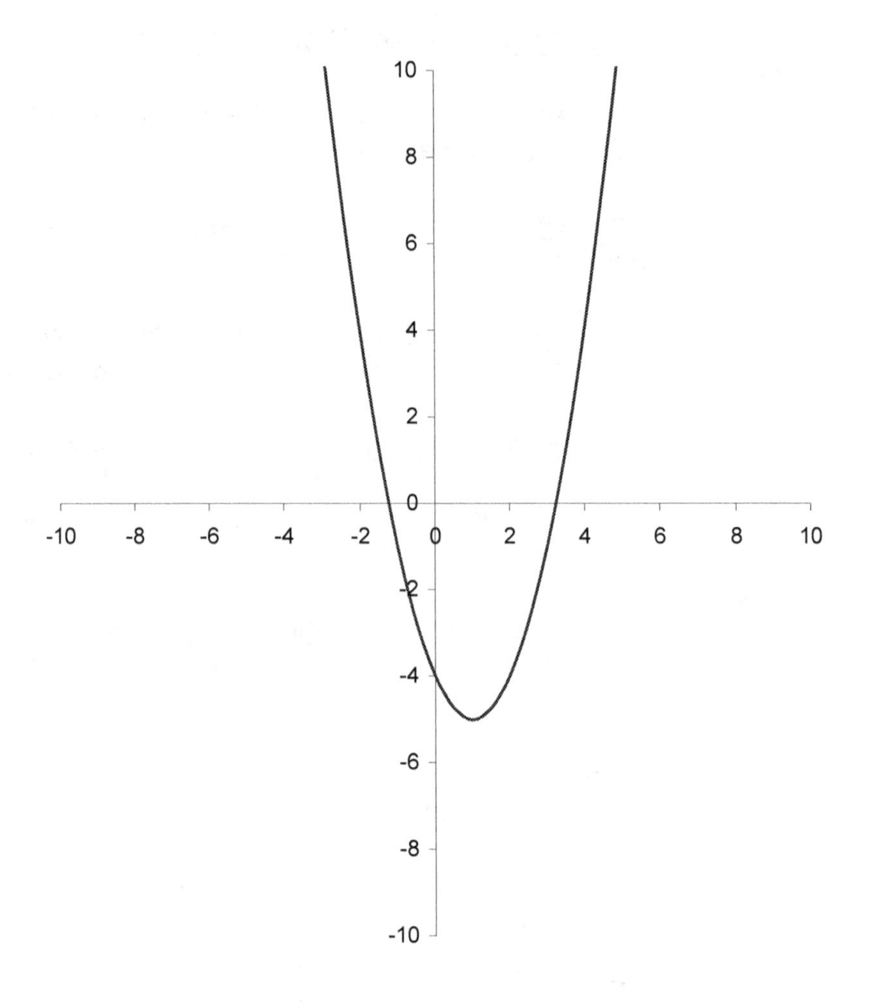

4. If $A = \{(x, y) \mid y = x^2 - 6\}$, find the domain and range.

5. Give the domain and range of set B if:

$$B = \{(1, ^-2), (4, ^-2), (7, ^-2), (6, ^-2)\}$$

6. Determine the domain of this function:

$$f(x) = \frac{5x + 7}{x^2 - 4}$$

7. Determine the domain and range of these graphs.

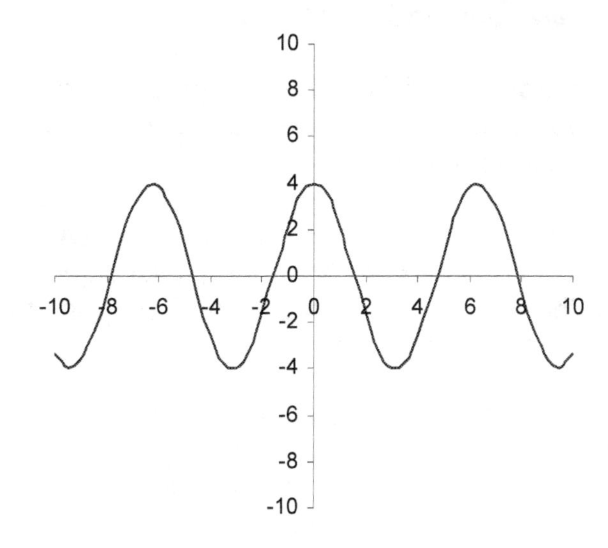

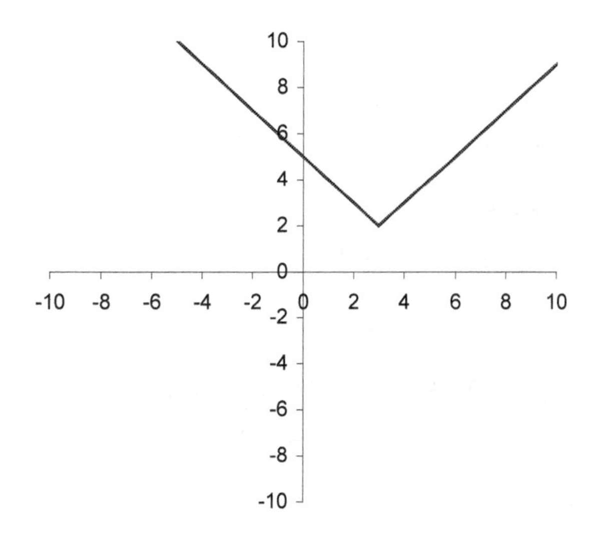

8. If $E = \{(x, y) \mid y = 5\}$, find the domain and range.

9. Determine the ordered pairs in the relation shown in this mapping.

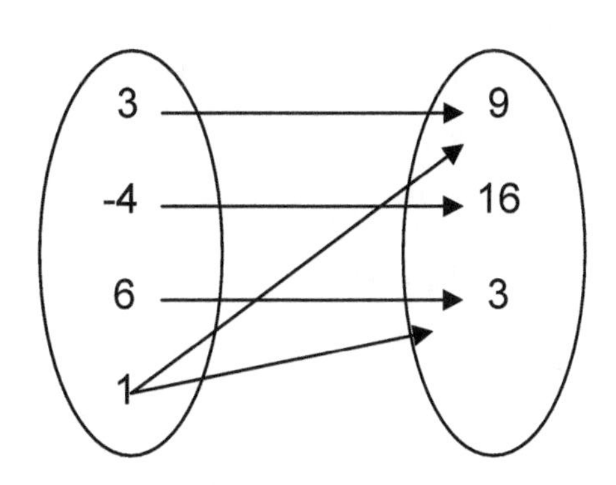

Competency 0008 Understand properties and applications of linear relations and functions.

When **graphing a first-degree equation**, solve for the variable. The graph of this solution will be a single point on the number line. There will be no arrows.

When graphing a linear inequality, the dot will be hollow if the inequality sign is < or >. If the inequality signs is either $\geq$ or $\leq$, the dot on the graph will be solid. The arrow goes to the right for $\geq$ or >. The arrow goes to the left for < or $\leq$.

Example:

$$5(x+2)+2x=3(x-2)$$
$$5x+10+2x=3x-6$$
$$7x+10=3x-6$$
$$4x=-16$$
$$x=-4$$

-10 ⁻8 ⁻6 ⁻4 ⁻2 **0** 2 4 6 8 10

Example:

$$2(3x-7)>10x-2$$
$$6x-14>10x-2$$
$$-4x>12$$
$$x<-3$$

Practice Problems:

1. $5x-1>14$
2. $7(2x-3)+5x=19-x$
3. $3x+42\geq 12x-12$
4. $5-4(x+3)=9$

A first degree equation has an equation of the form $ax + by = c$. To graph this equation, find both one point and the slope of the line or find two points. To find a point and slope, solve the equation for y. This gets the equation in **slope intercept form**, $y = mx + b$. The point (0, b) is the y-intercept and m is the line's slope. To find any two points, substitute any two numbers for x and solve for y. To find the intercepts, substitute 0 for x and then 0 for y.

Remember that graphs will go up as they go to the right when the slope is positive. Negative slopes make the lines go down as they go to the right.

If the equation solves to **x = any number**, then the graph is a **vertical line**. It only has an x intercept. Its slope is **undefined**.

If the equation solves to **y = any number**, then the graph is a **horizontal line**. It only has a y intercept. Its slope is 0 (zero).

When graphing a linear inequality, the line will be dotted if the inequality sign is $<$ or $>$. If the inequality signs are either $\geq$ or $\leq$, the line on the graph will be a solid line. Shade above the line when the inequality sign is $\geq$ or $>$. Shade below the line when the inequality sign is $<$ or $\leq$. Inequalities of the form $x >, x \leq, x <,$ or $x \geq$ number, draw a vertical line (solid or dotted). Shade to the right for $>$ or $\geq$. Shade to the left for $<$ or $\leq$. Remember: **Dividing or multiplying by a negative number will reverse the direction of the inequality sign.**

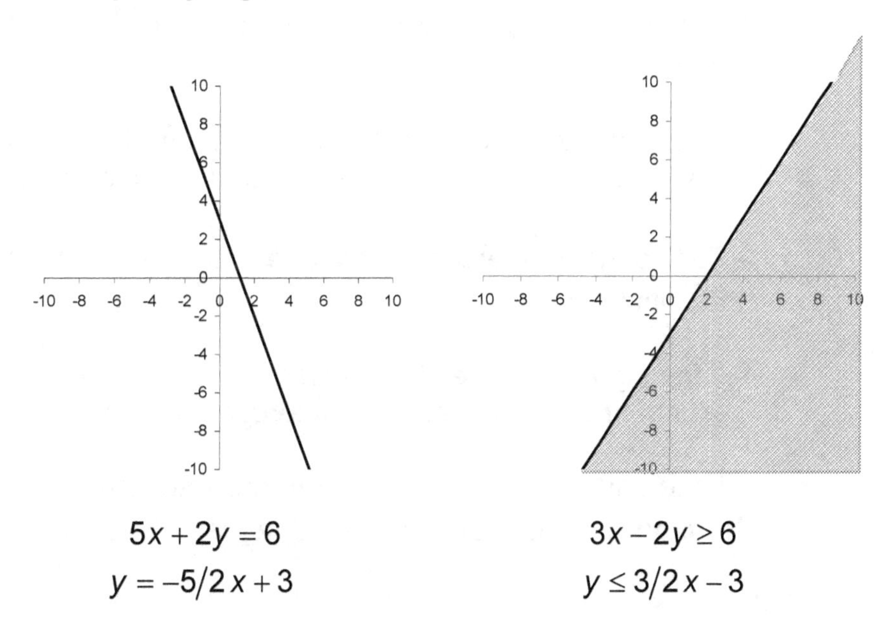

$5x + 2y = 6$
$y = -5/2x + 3$

$3x - 2y \geq 6$
$y \leq 3/2x - 3$

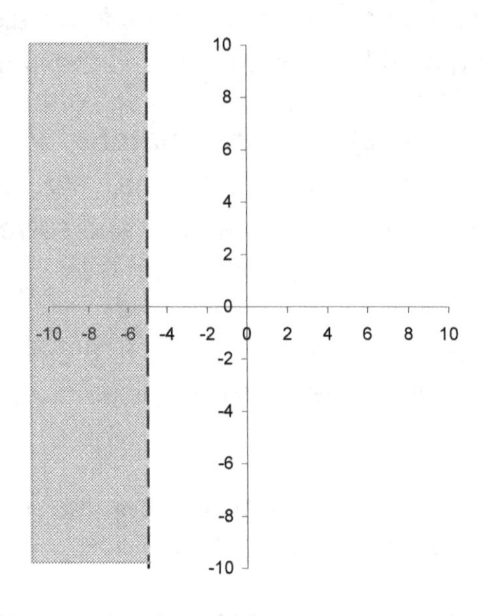

$$3x + 12 < -3$$
$$x < -5$$

Graph the following:

1. $2x - y = -4$
2. $x + 3y > 6$
3. $3x + 2y \leq 2y - 6$

- A **first degree equation** has an equation of the form $ax + by = c$. To find the slope of a line, solve the equation for y. This gets the equation into **slope intercept form**, $y = mx + b$. m is the line's slope.

- To find the y intercept, substitute 0 for x and solve for y. This is the y intercept. The y intercept is also the value of b in $y = mx + b$.

- To find the x intercept, substitute 0 for y and solve for x. This is the x intercept.

- If the equation solves to **x = any number**, then the graph is a **vertical line**. It only has an x intercept. Its slope is **undefined**.

- If the equation solves to **y = any number**, then the graph is a **horizontal line**. It only has a y intercept. Its slope is 0 (zero).

1. Find the slope and intercepts of $3x + 2y = 14$.

$$3x + 2y = 14$$
$$2y = {}^{-}3x + 14$$
$$y = {}^{-}2/3\ x + 7$$

The slope of the line is ${}^{-}2/3$, the value of m.
The y intercept of the line is 7.

The intercepts can also be found by substituting 0 in place of the other variable in the equation.

To find the y intercept:	To find the x intercept:
let $x = 0$; $3(0) + 2y = 14$	let $y = 0$; $3x + 2(0) = 14$
$0 + 2y = 14$	$3x + 0 = 14$
$2y = 14$	$3x = 14$
$y = 7$	$x = 14/3$
$(0,7)$ is the y intercept.	$(14/3,0)$ is the x intercept.

Find the slope and the intercepts (if they exist) for these equations:

1. $5x + 7y = {}^{-}70$
2. $x - 2y = 14$
3. $5x + 3y = 3(5 + y)$
4. $2x + 5y = 15$

- The **equation of a graph** can be found by finding its slope and its y intercept. To find the slope, find 2 points on the graph where co-ordinates are integer values. Using points: (x_1, y_1) and (x_2, y_2).

$$\text{slope} = \frac{y_2 - y_1}{x_2 - x_1}$$

The y intercept is the y coordinate of the point where a line crosses the y axis. The equation can be written in slope-intercept form, which is $y = mx + b$, where m is the slope and b is the y intercept. To rewrite the equation into some other form, multiply each term by the common denominator of all the fractions. Then rearrange terms as necessary.

- If the graph is a **vertical line**, then the equation solves to ***x* = the *x* co-ordinate of any point on the line**.

- If the graph is a **horizontal line**, then the equation solves to ***y* = the *y* coordinate of any point on the line**.

- **Given two points** on a line, the first thing to do is to find the slope of the line. If 2 points on the graph are (x_1, y_1) and (x_2, y_2), then the slope is found using the formula:

$$\text{slope} = \frac{y_2 - y_1}{x_2 - x_1}$$

The slope will now be denoted by the letter **m**. To write the equation of a line, choose either point. Substitute them into the formula:

$$Y - y_a = \text{m}(X - x_a)$$

Remember (x_a, y_a) can be (x_1, y_1) or (x_2, y_2) If **m**, the value of the slope, is distributed through the parentheses, the equation can be rewritten into other forms of the equation of a line.

Find the equation of a line through $(9, {}^-6)$ and $({}^-1, 2)$.

$$\text{slope} = \frac{y_2 - y_1}{x_2 - x_1} = \frac{2 - {}^-6}{{}^-1 - 9} = \frac{8}{{}^-10} = -\frac{4}{5}$$

$$Y - y_a = \text{m}(X - x_a) \rightarrow Y - 2 = {}^-4/5(X - {}^-1) \rightarrow$$
$$Y - 2 = {}^-4/5(X + 1) \rightarrow Y - 2 = {}^-4/5\,X - 4/5 \rightarrow$$
$$Y = {}^-4/5\,X + 6/5$$ This is the slope-intercept form.

Multiplying by 5 to eliminate fractions, it is:

$$5Y = {}^-4X + 6 \rightarrow 4X + 5Y = 6$$ Standard form.

Practice Problems:

Write the equation of a line through these two points:

1. $(5,8)$ and $({}^-3,2)$
2. $(11,10)$ and $(11,{}^-3)$
3. $({}^-4,6)$ and $(6,12)$
4. $(7,5)$ and $({}^-3,5)$

To graph an inequality, solve the inequality for y. This gets the inequality in **slope intercept form**, (for example $y < mx + b$). The point (0, b) is the y-intercept and m is the line's slope.

If the inequality solves to $x \geq$ **any number**, then the graph includes a **vertical line**.

If the inequality solves to $y \leq$ **any number**, then the graph includes a **horizontal line**.

When graphing a linear inequality, the line will be dotted if the inequality sign is $<$ or $>$. If the inequality signs are either $\geq$ or $\leq$, the line on the graph will be a solid line. Shade above the line when the inequality sign is $\geq$ or $>$. Shade below the line when the inequality sign is $<$ or $\leq$. For inequalities of the forms $x >$ number, $x \leq$ number, $x <$ number, or $x \geq$ number, draw a vertical line (solid or dotted). Shade to the right for $>$ or $\geq$. Shade to the left for $<$ or $\leq$.

Use these rules to graph and shade each inequality. The solution to a system of linear inequalities consists of the part of the graph that is shaded for each inequality. For instance, if the graph of one inequality is shaded with red, and the graph of another inequality is shaded with blue, then the overlapping area would be shaded purple. The purple area would be the points in the solution set of this system.

Example: Solve by graphing:

$$x + y \leq 6$$
$$x - 2y \leq 6$$

Solving the inequalities for y, they become:

$y \leq -x + 6$ (y-intercept of 6 and slope = -1)
$y \geq 1/2x - 3$ (y intercept of -3 and slope = $1/2$)

A graph with shading is shown below:

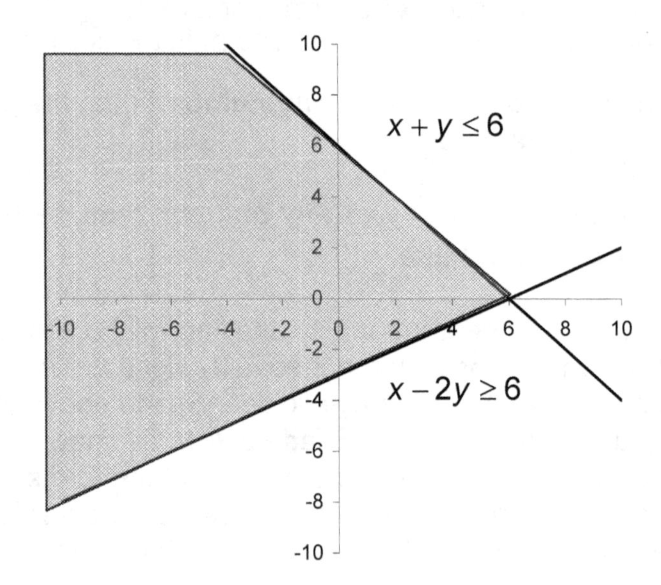

To solve an **equation or inequality**, follow these steps:

STEP 1. If there are parentheses, use the distributive property to eliminate them.

STEP 2. If there are fractions, determine their LCD (least common denominator). Multiply every term of the equation by the LCD. This will cancel out all of the fractions while solving the equation or inequality.

STEP 3. If there are decimals, find the largest decimal. Multiply each term by a power of 10(10, 100, 1000,etc.) with the same number of zeros as the length of the decimal. This will eliminate all decimals while solving the equation or inequality.

STEP 4. Combine like terms on each side of the equation or inequality.

STEP 5. If there are variables on both sides of the equation, add or subtract one of those variable terms to move it to the other side. Combine like terms.

STEP 6. If there are constants on both sides, add or subtract one of those constants to move it to the other side. Combine like terms.

STEP 7. If there is a coefficient in front of the variable, divide both sides by this number. This is the answer to an equation. However, remember:

Dividing or multiplying an inequality by a negative number will reverse the direction of the inequality sign.

STEP 8. The solution of a linear equation solves to one single number. The solution of an inequality is always stated including the inequality sign.

Example: Solve: $3(2x+5)-4x=5(x+9)$

$6x+15-4x=5x+45$	ref. step 1
$2x+15=5x+45$	ref. step 4
$^-3x+15=45$	ref. step 5
$^-3x=30$	ref. step 6
$x={}^-10$	ref. step 7

Example: Solve: $1/2(5x+34)=1/4(3x-5)$

$5/2x+17=3/4x-5/4$	ref.step 1
LCD of 5/2, 3/4, and 5/4 is 4.	
Multiply by the LCD of 4.	
$4(5/2x+17)=(3/4x-5/4)4$	ref.step 2
$10x+68=3x-5$	
$7x+68={}^-5$	ref.step 5
$7x={}^-73$	ref.step 6
$x={}^-73/7$ or $^-10\ 3/7$	ref.step 7

Check:

$$\frac{1}{2}\left[5\frac{-13}{7}+34\right]=\frac{1}{4}\left[3\left(\frac{-13}{7}\right)-\frac{5}{4}\right]$$

$$\frac{1}{2}\left[\frac{-13(5)}{7}+34\right]=\frac{1}{4}\left[3\left(\frac{-13}{7}\right)-\frac{5}{4}\right]$$

$$\frac{-13(5)}{7}+17=\frac{3(-13)}{28}-\frac{5}{4}$$

$$\frac{-13(5)+17(14)}{14}=\frac{3(-13)}{28}-\frac{5}{4}$$

$$\left[-13(5)+17(14)\right]2=\frac{3(-13)-35}{28}$$

$$\frac{-130+476}{28}=\frac{-219-35}{28}$$

$$\frac{-254}{28}=\frac{-254}{28}$$

Example: Solve: $6x+21<8x+31$

$^{-}2x+21<31$ ref. step 5

$^{-}2x<10$ ref. step 6

$x>{}^{-}5$ ref. step 7

Note that the inequality sign has changed.

Equations and inequalities can be used to solve various types of word problems.

Example: The YMCA wants to sell raffle tickets to raise at least $32,000. If they must pay $7,250 in expenses and prizes out of the money collected from the tickets, how many tickets worth $25 each must they sell?

Solution: Since they want to raise **at least $32,000**, that means they would be happy to get 32,000 **or more**. This requires an inequality.

Let $x =$ number of tickets sold
Then $25x =$ total money collected for x tickets

Total money minus expenses is greater than $32,000.

$$25x - 7250 \geq 32000$$
$$25x \geq 39250$$
$$x \geq 1570$$

If they sell **1,570 tickets or more**, they will raise AT LEAST $32,000.

Example: The Simpsons went out for dinner. All four of them ordered the aardvark steak dinner. Bert paid for the 4 meals and included a tip of $12 for a total of $84.60. How much was an aardvark steak dinner?

Let $x =$ the price of one aardvark dinner.
So $4x =$ the price of 4 aardvark dinners.

$$4x + 12 = 84.60$$
$$4x = 72.60$$
$$x = \$18.15 \text{ for each dinner.}$$

Some word problems can be solved using a system of equations or inequalities. Watch for words like greater than, less than, at least, or no more than, which indicate the need for inequalities.

Some word problems can be solved using a system of equations or inequalities. Watch for words like greater than, less than, at least, or no more than which indicate the need for inequalities.

Example: Farmer Greenjeans bought 4 cows and 6 sheep for $1700. Mr. Ziffel bought 3 cows and 12 sheep for $2400. If all the cows were the same price and all the sheep were another price, find the price charged for a cow or for a sheep.

Let $x =$ price of a cow
Let $y =$ price of a sheep

Then Farmer Greenjeans' equation would be: $4x + 6y = 1700$
Mr. Ziffel's equation would be: $3x + 12y = 2400$

To solve by **addition-subtraction**:

Multiply the first equation by $^{-}2$: $^{-}2(4x + 6y = 1700)$
Keep the other equation the same : $(3x + 12y = 2400)$
By doing this, the equations can be added to each other to eliminate one variable and solve for the other variable.

$$\begin{aligned} ^{-}8x - 12y &= {}^{-}3400 \\ 3x + 12y &= 2400 \\ \hline ^{-}5x \qquad &= {}^{-}1000 \end{aligned}$$

Add these equations.

$x = 200 \leftarrow$ the price of a cow was $200.
Solving for y, $y = 150 \leftarrow$ the price of a sheep,$150.

(This problem can also be solved by substitution or determinants.)

Example: John has 9 coins, which are either dimes or nickels, that are worth $.65. Determine how many of each coin he has.

Let $d =$ number of dimes.
Let $n =$ number of nickels.
The number of coins total 9.
The value of the coins equals 65.

Then: $n + d = 9$
$5n + 10d = 65$

Multiplying the first equation by $^{-}5$, it becomes:

$$^{-}5n - 5d = {}^{-}45$$
$$\underline{5n + 10d = 65}$$
$$5d = 20$$

$d = 4$ There are 4 dimes, so there are $(9 - 4)$ or 5 nickels.

Example: Sharon's Bike Shoppe can assemble a 3 speed bike in 30 minutes or a 10 speed bike in 60 minutes. The profit on each bike sold is $60 for a 3 speed or $75 for a 10 speed bike. How many of each type of bike should they assemble during an 8 hour day (480 minutes) to make the maximum profit? Total daily profit must be at least $300.

Let $x =$ number of 3 speed bikes.
$y =$ number of 10 speed bikes.

Since there are only 480 minutes to use each day,

$30x + 60y \leq 480$ is the first inequality.

Since the total daily profit must be at least $300,

$60x + 75y \geq 300$ is the second inequality.

$30x + 60y \leq 480$ solves to $y \leq 8 - 1/2x$

$$60y \leq -30x + 480$$

$$y \leq -\frac{1}{2}x + 8$$

$60x + 75y \geq 300$ solves to $y \geq 4 - 4/5x$

$$75y + 60x \geq 300$$

$$75y \geq -60x + 300$$

$$y \geq -\frac{4}{5}x + 4$$

Graph these 2 inequalities:

$y \le 8 - 1/2x$
$y \ge 4 - 4/5x$

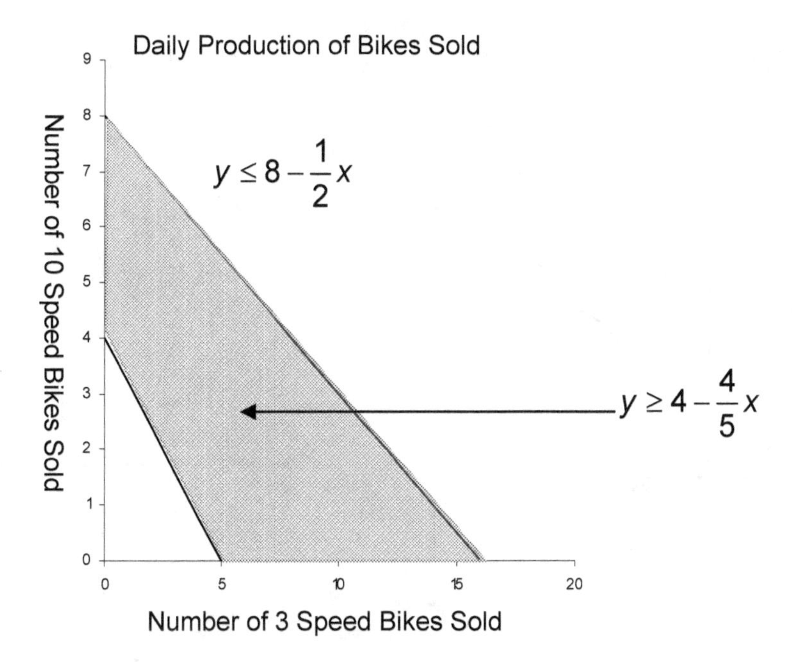

Realize that $x \ge 0$ and $y \ge 0$, since the number of bikes assembled can not be a negative number. Graph these as additional constraints on the problem. The number of bikes assembled must always be an integer value, so points within the shaded area of the graph must have integer values. The maximum profit will occur at or near a corner of the shaded portion of this graph. Those points occur at (0,4), (0,8), (16,0), or (5,0).
Since profits are $60/3-speed or $75/10-speed, the profit would be :

(0,4)	$60(0)+75(4)=300$
(0,8)	$60(0)+75(8)=600$
(16,0)	$60(16)+75(0)=960 \leftarrow$ Maximum profit
(5,0)	$60(5)+75(0)=300$

The maximum profit would occur if 16 3-speed bikes are made daily.

Functions defined by two or more formulas are **piecewise functions**. The formula used to evaluate piecewise functions varies depending on the value of x. The graphs of piecewise functions consist of two or more pieces, or intervals, and are often discontinuous.

Example 1:

$$f(x) = \begin{cases} x + 1 & \text{if } x > 2 \\ x - 2 & \text{if } x \leq 2 \end{cases}$$

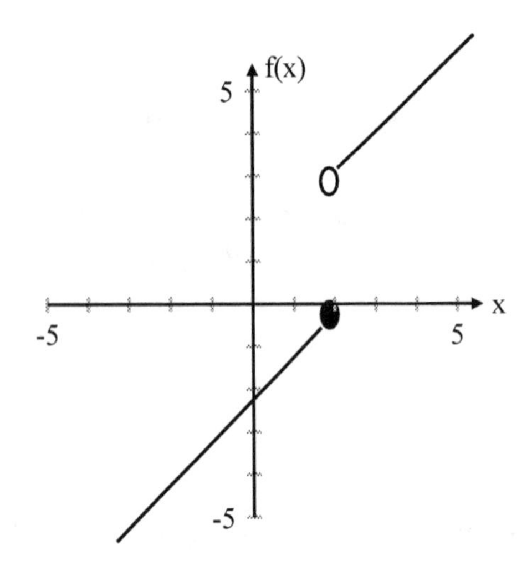

Example 2:

$$f(x) = \begin{cases} x & \text{if } x \geq 1 \\ x^2 & \text{if } x < 1 \end{cases}$$

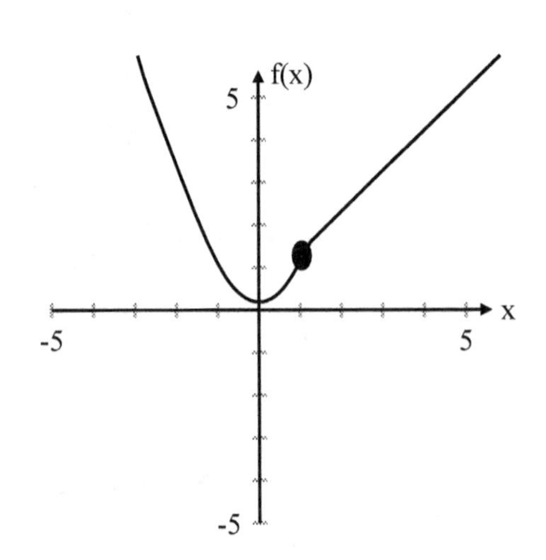

When graphing or interpreting the graph of piecewise functions it is important to note the points at the beginning and end of each interval because the graph must clearly indicate what happens at the end of each interval. Note that in the graph of Example 1, point (2, 3) is not part of the graph and is represented by an empty circle. On the other hand, point (2, 0) is part of the graph and is represented as a solid circle. Note also that the graph of Example 2 is continuous despite representing a piecewise function.

Practice: Graph the following piecewise equations.

1. $f(x) = x^2$ if $x > 0$
   $= x + 4$ if $x \leq 0$

2. $f(x) = x^2 - 1$ if $x > 2$
   $= x^2 + 2$ if $x \leq 2$

Composition is a process that creates a new function by substituting an entire function into another function. The composition of two functions f(x) and g(x) is denoted by $(f \circ g)(x)$ or f(g(x)). The domain of the composed function, f(g(x)), is the set of all values of x in the domain of g that produce a value for g(x) that is in the domain of f. In other words, f(g(x)) is defined whenever both g(x) and f(g(x)) are defined.

Example 1:

If f(x) = x + 1 and $g(x) = x^3$, find the composition functions $f \circ g$ and $g \circ f$ and state their domains.

Solution:

$(f \circ g)(x) = f(g(x)) = f(x^3) = x^3 + 1$
$(g \circ f)(x) = g(f(x)) = g(x + 1) = (x + 1)^3$

The domain of both composite functions is the set of all real numbers.

Note that f(g(x)) and g(f(x)) are not the same. In general, unlike multiplication and addition, composition is not reversible. Thus, the order of composition is important.
Example 2:

If f(x) = sqrt(x) and g(x) = x +2, find the composition functions $f \circ g$ and $g \circ f$ and state their domains.

Solution:

$(f \circ g)(x)$ = f(g(x)) = f(x + 2) = sqrt(x + 2)
$(g \circ f)(x)$ = g(f(x)) = g(sqrt(x)) = sqrt(x) +2

The domain of f(g(x)) is $x \geq -2$ because x + 2 must be non-negative in order to take the square root.

The domain of g(f(x)) is $x \geq 0$ because x must be non-negative in order to take the square root.Note that defining the domain of composite functions is important when square roots are involved.

Competency 0009 Understand properties and applications of quadratic relations and functions.

A **quadratic equation** is written in the form $ax^2 + bx + c = 0$. To solve a quadratic equation by factoring, at least one of the factors must equal zero.

Example:
Solve the equation.

$x^2 + 10x - 24 = 0$	
$(x + 12)(x - 2) = 0$	Factor.
$x + 12 = 0$ or $x - 2 = 0$	Set each factor equal to 0.
$x = {}^{-}12 \qquad x = 2$	Solve.

Check:
$x^2 + 10x - 24 = 0$

$({}^{-}12)^2 + 10({}^{-}12) - 24 = 0$	$(2)^2 + 10(2) - 24 = 0$
$144 - 120 - 24 = 0$	$4 + 20 - 24 = 0$
$0 = 0$	$0 = 0$

A quadratic equation that cannot be solved by factoring can be solved by **completing the square**.

Example:

Solve the equation.

$x^2 - 6x + 8 = 0$	
$x^2 - 6x = {}^{-}8$	Move the constant to the right side.
$x^2 - 6x + 9 = {}^{-}8 + 9$	Add the square of half the coefficient of x to both sides.
$(x - 3)^2 = 1$	Write the left side as a perfect square.
$x - 3 = \pm\sqrt{1}$	Take the square root of both sides.
$x - 3 = 1 \qquad x - 3 = {}^{-}1$	Solve.
$x = 4 \qquad x = 2$	

Check:

$x^2 - 6x + 8 = 0$

$4^2 - 6(4) + 8 = 0$	$2^2 - 6(2) + 8 = 0$
$16 - 24 + 8 = 0$	$4 - 12 + 8 = 0$
$0 = 0$	$0 = 0$

The general technique for graphing quadratics is the same as for graphing linear equations. Graphing quadratic equations, however, results in a parabola instead of a straight line.

To solve a quadratic equation using the **quadratic formula**, be sure that your equation is in the form $ax^2 + bx + c = 0$. Substitute these values into the formula:

$$x = \frac{-b \pm \sqrt{b^2 - 4ac}}{2a}$$

Example:

Graph $y = 3x^2 + x - 2$.

x	$y = 3x^2 + x - 2$
-2	8
-1	0
0	-2
1	2
2	12

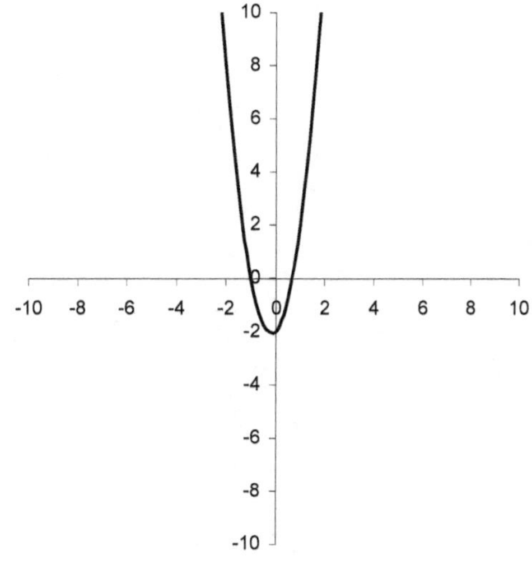

Example:

Solve the equation.

$$3x^2 = 7 + 2x \rightarrow 3x^2 - 2x - 7 = 0$$

$$a = 3 \quad b = {}^-2 \quad c = {}^-7$$

$$x = \frac{-({}^-2) \pm \sqrt{({}^-2)^2 - 4(3)({}^-7)}}{2(3)}$$

$$x = \frac{2 \pm \sqrt{4 + 84}}{6}$$

$$x = \frac{2 \pm \sqrt{88}}{6}$$

$$x = \frac{2 \pm 2\sqrt{22}}{6}$$

$$x = \frac{1 \pm \sqrt{22}}{3}$$

Some word problems will give a quadratic equation to be solved. When the quadratic equation is found, set it equal to zero and solve the equation by factoring or the quadratic formula. Examples of this type of problem follow.

Example:
Ashland (A) is a certain distance north of Belmont (B). The distance from Belmont east to Carlisle (C) is 5 miles more than the distance from Ashland to Belmont. The distance from Ashland to Carlisle is 10 miles more than the distance from Alberta to Belmont. How far is Ashland from Carlisle?

Solution:
Since north and east form a right angle, these distances are the lengths of the legs of a right triangle. If the distance from Ashland to Belmont is x, then from Belmont to Carlisle is $x+5$, and the distance from Ashland to Carlisle is $x+10$.

The equation is: $AB^2+BC^2=AC^2$

$x^2+(x+5)^2=(x+10)^2$

$x^2+x^2+10x+25=x^2+20x+100$

$2x^2+10x+25=x^2+20x+100$

$x^2-10x-75=0$

$(x-15)(x+5)=0$ Distance cannot be negative.

$x=15$ Distance from Ashland to Belmont.

$x+5=20$ Distance from Belmont to Carlisle.

$x+10=25$ Distance from Ashland to Carlisle.

Example:
The square of a number is equal to 6 more than the original number. Find the original number.

Solution: If $x=$ original number, then the equation is:

$x^2=6+x$ Set this equal to zero.

$x^2-x-6=0$ Now factor.

$(x-3)(x+2)=0$

$x=3$ or $x={}^-2$ There are 2 solutions, 3 or $^-2$.

Try these:

1. One side of a right triangle is 1 less than twice the shortest side, while the third side of the triangle is 1 more than twice the shortest side. Find all 3 sides. (changed font)

2. Twice the square of a number equals 2 less than 5 times the number. Find the number(s).

The **discriminant** is the portion of the quadratic formula which is found under the square root sign; that is $b^2 - 4ac$.

The discriminant can be used to determine the nature of the solution of a quadratic equation.

There are three possibilities:
$b^2 - 4ac > 0$, then there are 2 real roots
$b^2 - 4ac = 0$, then exactly one real root exists
$b^2 - 4ac < 0$. then there are no real roots

Follow these steps to write a quadratic equation from its roots:

1. Add the roots together. The answer is their **sum**. Multiply the roots together. The answer is their **product**.
2. A quadratic equation can be written using the sum and product like this:

$$x^2 + (\textbf{opposite of the sum})x + \textbf{product} = 0$$

3. If there are any fractions in the equation, multiply every term by the common denominator to eliminate the fractions. This is the quadratic equation.
4. If a quadratic equation has only 1 root, use it twice and follow the first 3 steps above.

Example:
Find a quadratic equation with roots of 4 and $^{-}9$.

Solutions:
The sum of 4 and $^{-}9$ is $^{-}5$. The product of 4 and $^{-}9$ is $^{-}36$.
The equation would be:

$$\mathbf{x^2 + (\text{opposite of the sum})x + \text{product} = 0}$$
$$\mathbf{x^2 + 5x - 36 = 0}$$

Find a quadratic equation with roots of $5+2i$ and $5-2i$.

Solutions:
The sum of $5+2i$ and $5-2i$ is 10. The product of $5+2i$ and $5-2i$ is $25-4i^2 = 25+4 = 29$.(font)

The equation would be:

$$\mathbf{x^2 + (\text{opposite of the sum})x + \text{product} = 0}$$
$$\mathbf{x^2 - 10x + 29 = 0}$$

Find a quadratic equation with roots of $2/3$ and $^{-}3/4$.

Solutions:
The sum of $2/3$ and $^{-}3/4$ is $^{-}1/12$. The product of $2/3$ and $^{-}3/4$ is $^{-}1/2$.

The equation would be :

$$\mathbf{x^2 + (\text{opposite of the sum})x + \text{product} = 0}$$
$$\mathbf{x^2 + 1/12\,x - 1/2 = 0}$$

Common denominator = 12, so multiply by 12.

$$\mathbf{12(x^2 + 1/12\,x - 1/2 = 0}$$
$$\mathbf{12x^2 + 1x - 6 = 0}$$
$$\mathbf{12x^2 + x - 6 = 0}$$

Try these:
1. Find a quadratic equation with a root of 5.
2. Find a quadratic equation with roots of $8/5$ and $^{-}6/5$.
3. Find a quadratic equation with roots of 12 and $^{-}3$.

Some word problems can be solved by setting up a quadratic equation or inequality. Examples of this type could be problems that deal with finding a maximum area.

Example:

A family wants to enclose 3 sides of a rectangular garden with 200 feet of fence. In order to have a garden with an area of **at least** 4800 square feet, find the dimensions of the garden. Assume that a wall or a fence already borders the fourth side of the garden

Solution:
Let $x =$ distance from the wall

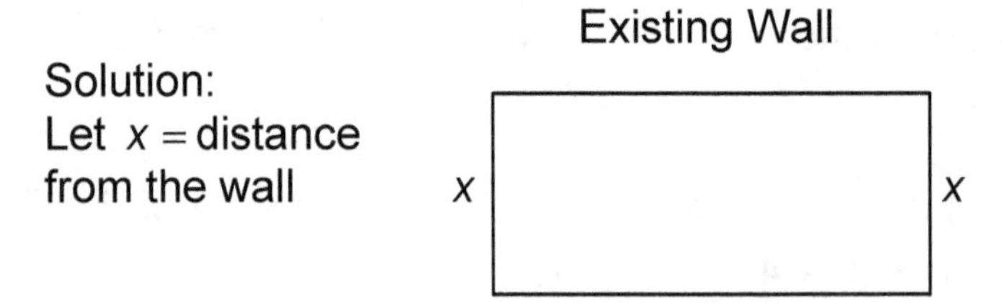

Then 2x feet of fence is used for these 2 sides. The remaining side of the garden would use the rest of the 200 feet of fence, that is, $200-2x$ feet of fence. Therefore the width of the garden is x feet and the length is $200-2x$ ft.

The area, $200x-2x^2$, needs to be greater than or equal to 4800 sq. ft. So, this problem uses the inequality $4800 \leq 200x-2x^2$. This becomes $2x^2-200x+4800 \leq 0$. Solving this, we get:

$$200x - 2x^2 \geq 4800$$
$$-2x^2 + 200x - 4800 \geq 0$$
$$2\left(-x^2 + 100x - 2400\right) \geq 0$$
$$-x^2 + 100x - 2400 \geq 0$$
$$(-x+60)(x-40) \geq 0$$
$$-x + 60 \geq 0$$
$$-x \geq -60$$
$$x \leq 60$$
$$x - 40 \geq 0$$
$$x \geq 40$$

So the area will be at least 4800 square feet if the width of the garden is from 40 up to 60 feet. (The length of the rectangle would vary from 120 feet to 80 feet depending on the width of the garden.)

Quadratic equations can be used to model **different real life situations**. The graphs of these quadratics can be used to determine information about this real life situation.

Example:

The height of a projectile fired upward at a velocity of v meters per second from an original height of h meters is $y = h + vx - 4.9x^2$. If a rocket is fired from an original height of 250 meters with an original velocity of 4800 meters per second, find the approximate time the rocket would drop to sea level (a height of 0).

Solution:
The equation for this problem is: $y = 250 + 4800x - 4.9x^2$. If the height at sea level is zero, then $y = 0$ so $0 = 250 + 4800x - 4.9x^2$. Solving this for x could be done by using the quadratic formula. In addition, the approximate time in seconds (x) until the rocket would be at sea level could be estimated by looking at the graph. When the y value of the graph goes from positive to negative then there is a root (also called solution or x intercept) in that interval.

$$x = \frac{^{-}4800 \pm \sqrt{4800^2 - 4(^{-}4.9)(250)}}{2(^{-}4.9)} \approx 980 \text{ or } ^{-}0.05 \text{ seconds}$$

Since the time has to be positive, it will be about 980 seconds until the rocket is at sea level.

To graph an **inequality**, graph the quadratic as if it were an equation; however, if the inequality has just a $>$ or $<$ sign, then make the curve itself dotted. Shade above the curve for $>$ or $\geq$. Shade below the curve for $<$ or $\leq$.

Examples:

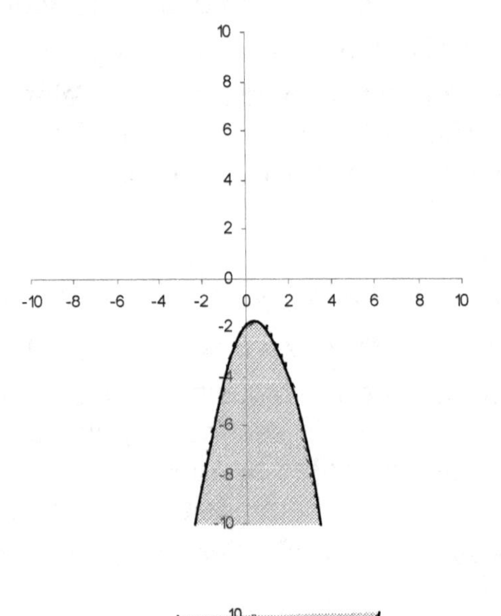

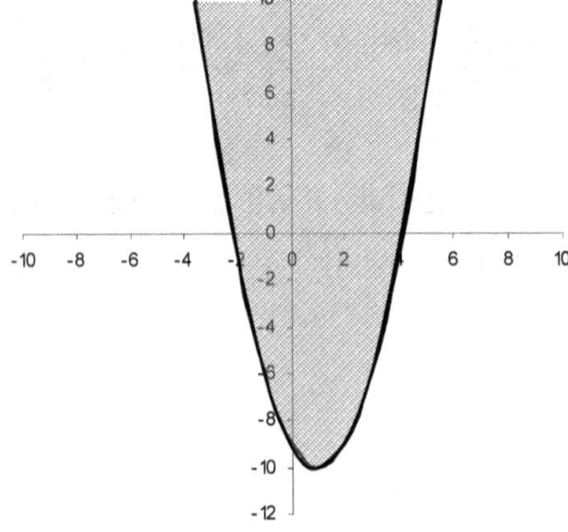

Competency 0010 Understand properties and applications of exponential, polynomial, rational, and absolute value functions and relations.

Some problems can be solved using equations with rational expressions. First write the equation. To solve it, multiply each term by the LCD of all fractions. This will cancel out all of the denominators and give an equivalent algebraic equation that can be solved.

1. The denominator of a fraction is two less than three times the numerator. If 3 is added to both the numerator and denominator, the new fraction equals $1/2$.

original fraction: $\frac{x}{3x-2}$ revised fraction: $\frac{x+3}{3x+1}$

$$\frac{x+3}{3x+1}=\frac{1}{2}$$

$$2x+6=3x+1$$

$$x=5$$

original fraction: $\frac{5}{13}$

2. Elly Mae can feed the animals in 15 minutes. Jethro can feed them in 10 minutes. How long will it take them if they work together?

Solution: If Elly Mae can feed the animals in 15 minutes, then she could feed $1/15$ of them in 1 minute, $2/15$ of them in 2 minutes, $x/15$ of them in x minutes. In the same fashion Jethro could feed $x/10$ of them in x minutes. Together they complete 1 job. The equation is:

$$\frac{x}{15}+\frac{x}{10}=1$$

Multiply each term by the LCD of 30:

$$2x+3x=30$$

$$x=6 \text{ minutes}$$

3. A salesman drove 480 miles from Pittsburgh to Hartford. The next day he returned the same distance to Pittsburgh in half an hour less time than his original trip took, because he increased his average speed by 4 mph. Find his original speed.

Since distance = rate x time then time = $\dfrac{\text{distance}}{\text{rate}}$

original time $-1/2$ hour $=$ shorter return time

$$\frac{480}{x}-\frac{1}{2}=\frac{480}{x+4}$$

Multiplying by the LCD of $2x(x+4)$, the equation becomes:

$480\left[2(x+4)\right]-1\left[x(x+4)\right]=480(2x)$

$960x+3840-x^2-4x=960x$

$x^2+4x-3840=0$

$(x+64)(x-60)=0$ Either (x-60=0) or (x+64=0) or both=0

$x=60$ 60 mph is the original speed.

This is the solution since the time cannot be negative. Check your answer

$x+4=64$

$$\frac{480}{60}-\frac{1}{2}=\frac{480}{64}$$

$$8-\frac{1}{2}=7\frac{1}{2}$$

$$7\frac{1}{2}=7\frac{1}{2}$$

Try these:

1. Working together, Larry, Moe, and Curly can paint an elephant in 3 minutes. Working alone, it would take Larry 10 minutes or Moe 6 minutes to paint the elephant. How long would it take Curly to paint the elephant if he worked alone?
2. The denominator of a fraction is 5 more than twice the numerator. If the numerator is doubled, and the denominator is increased by 5, the new fraction is equal to $1/2$. Find the original number.
3. The distance from Augusta, Maine to Galveston, Texas is 2108 miles. If a car drove 6 mph faster than a truck and got to Galveston 3 hours before the truck, find the speeds of the car and the truck.

There are 2 easy ways to find the values of a function. First, to find the value of a function when $x = 3$, substitute 3 in place of every letter x. Then simplify the expression following the order of operations. For example, if $f(x) = x^3 - 6x + 4$, then to find f(3), substitute 3 for x.

The equation becomes $f(3) = 3^3 - 6(3) + 4 = 27 - 18 + 4 = 13$.
So (3, 13) is a point on the graph of f(x).

A second way to find the value of a function is to use synthetic division. To find the value of a function when $x = 3$, divide 3 into the coefficients of the function. (Remember that coefficients of missing terms, like x^2 , must be included). The remainder is the value of the function.
If $f(x) = x^3 - 6x + 4$, then to find f(3) using synthetic division:

Note the 0 for the missing x^2 term.

$$\begin{array}{r|rrrr} 3 & 1 & 0 & ^{-}6 & 4 \\ & & 3 & 9 & 9 \\ \hline & 1 & 3 & 3 & 13 \end{array}$$

$13 \leftarrow$ this is the value of the function.

Therefore, (3, 13) is a point on the graph of $f(x) = x^3 - 6x + 4$.

Example: Find values of the function at integer values from x = -3 to x = 3 if $f(x) = x^3 - 6x + 4$.

If $x = {}^-3$:

$$f({}^-3) = ({}^-3)^3 - 6({}^-3) + 4$$
$$= ({}^-27) - 6({}^-3) + 4$$
$$= {}^-27 + 18 + 4 = {}^-5$$

Using synthetic division:

$$\begin{array}{r|rrrr} {}^-3 & 1 & 0 & {}^-6 & 4 \\ & & {}^-3 & 9 & {}^-9 \\ \hline & 1 & {}^-3 & 3 & {}^-5 \end{array}$$

$1 \ {}^-3 \ 3 \ {}^-5 \leftarrow$ this is the value of the function if $x = {}^-3$.
Therefore, $({}^-3, {}^-5)$ is a point on the graph.

If $x = {}^-2$:

$$f({}^-2) = ({}^-2)^3 - 6({}^-2) + 4$$
$$= ({}^-8) - 6({}^-2) + 4$$

$= {}^-8 + 12 + 4 = 8 \leftarrow$ this is the value of the function if $x = {}^-2$.
Therefore, $({}^-2, 8)$ is a point on the graph.

If $x = {}^-1$:

$$f({}^-1) = ({}^-1)^3 - 6({}^-1) + 4$$
$$= ({}^-1) - 6({}^-1) + 4$$
$$= {}^-1 + 6 + 4 = 9$$

Using synthetic division:

$$\begin{array}{r|rrrr} {}^-1 & 1 & 0 & {}^-6 & 4 \\ & & {}^-1 & 1 & 5 \\ \hline & 1 & {}^-1 & {}^-5 & 9 \end{array}$$

$1 \ {}^-1 \ {}^-5 \ 9 \leftarrow$ this is the value if the function if $x = {}^-1$.
Therefore, $({}^-1, 9)$ is a point on the graph.

If $x = 0$:

$$\begin{aligned} f(0) &= (0)^3 - 6(0) + 4 \\ &= 0 - 6(0) + 4 \\ &= 0 - 0 + 4 = 4 \leftarrow \text{this is the value of the function if } x = 0. \end{aligned}$$

Therefore, (0, 4) is a point on the graph.

If $x = 1$:

$$\begin{aligned} f(1) &= (1)^3 - 6(1) + 4 \\ &= (1) - 6(1) + 4 \\ &= 1 - 6 + 4 = {}^-1 \end{aligned}$$

Using synthetic division:

$$\begin{array}{r|rrrr} 1 & 1 & 0 & {}^-6 & 4 \\ & & 1 & 1 & {}^-5 \\ \hline & 1 & 1 & {}^-5 & {}^-1 \end{array} \leftarrow \text{this is the value if the function of } x = 1.$$

Therefore, (1, $^-1$) is a point on the graph.

If $x = 2$:

$$\begin{aligned} f(2) &= (2)^3 - 6(2) + 4 \\ &= 8 - 6(2) + 4 \\ &= 8 - 12 + 4 = 0 \end{aligned}$$

Using synthetic division:

$$\begin{array}{r|rrrr} 2 & 1 & 0 & {}^-6 & 4 \\ & & 2 & 4 & {}^-4 \\ \hline & 1 & 2 & {}^-2 & 0 \end{array} \leftarrow \text{this is the value of the function if } x = 2.$$

Therefore, (2, 0) is a point on the graph.

If $x = 3$:

$$\begin{aligned} f(3) &= (3)^3 - 6(3) + 4 \\ &= 27 - 6(3) + 4 \\ &= 27 - 18 + 4 = 13 \end{aligned}$$

Using synthetic division:

$$\begin{array}{r|rrrr} 3 & 1 & 0 & ^-6 & 4 \\ & & 3 & 9 & 9 \\ \hline & 1 & 3 & 3 & 13 \end{array}$$

$\leftarrow$ this is the value of the function if $x = 3$.
Therefore, (3, 13) is a point on the graph.

The following points are points on the graph:

X	Y
$^-3$	$^-5$
$^-2$	8
$^-1$	9
0	4
1	$^-1$
2	0
3	13

Note the change in sign of the y value between $x = {}^-3$ and $x = {}^-2$. This indicates there is a zero between $x = {}^-3$ and $x = {}^-2$. Since there is another change in sign of the y value between $x = 0$ and $x = {}^-1$, there is a second root there. When $x = 2$, $y = 0$ so $x = 2$ is an exact root of this polynomial.

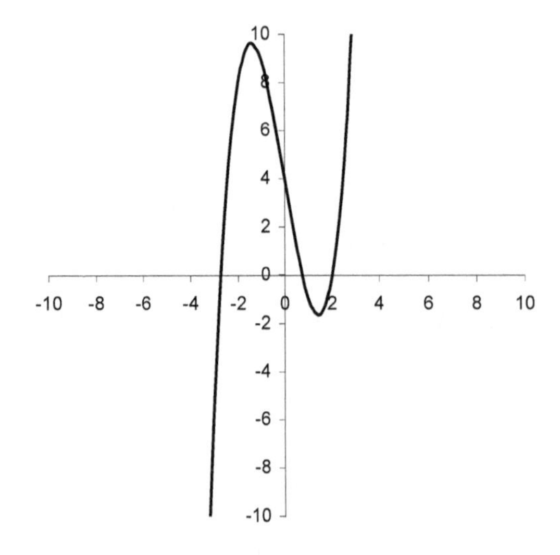

Word problems can sometimes be solved by using a system of two equations in two unknowns. This system can then be solved using substitution, the addition-subtraction method, or graphing.

Example: Mrs. Winters bought 4 dresses and 6 pairs of shoes for \$340. Mrs. Summers went to the same store and bought 3 dresses and 8 pairs of shoes for \$360. If all the dresses were the same price and all the shoes were the same price, find the price charged for a dress and for a pair of shoes.

Let x = price of a dress
Let y = price of a pair of shoes

Then Mrs. Winters' equation would be: $4x+6y=340$
Mrs. Summers' equation would be: $3x+8y=360$

To solve by addition-subtraction:

Multiply the first equation by 4: $4(4x+6y=340)$
Multiply the other equation by -3: $-3(3x+8y=360)$
By doing this, the equations can be added to each other to eliminate one variable and solve for the other variable.

$$\begin{array}{r} 16x+24y=1360 \\ \underline{-9x-24y=-1080} \\ 7x=280 \\ x=40 \end{array}$$

$x=40 \leftarrow$ the price of a dress was \$40

solving for y, $y=30$ $\leftarrow$ the price of a pair of shoes, \$30

Example: Aardvark Taxi charges \$4 initially plus \$1 for every mile traveled. Baboon Taxi charges \$6 initially plus \$.75 for every mile traveled. Determine when it is cheaper to ride with Aardvark Taxi and when to ride with Baboon Taxi.

Aardvark Taxi's equation: $y=1x+4$
Baboon Taxi's equation : $y=.75x+6$

Using substitution: $.75x+6=x+4$
Multiplying by 4: $3x+24=4x+16$
Solving for x : $8=x$

This tells you that at 8 miles the total charge for the companies is the same. Clearly, Aardvark is cheaper for distances up to 8 miles, but Baboon Taxi is cheaper for distances greater than 8 miles.

This problem can also be solved by graphing the 2 equations.

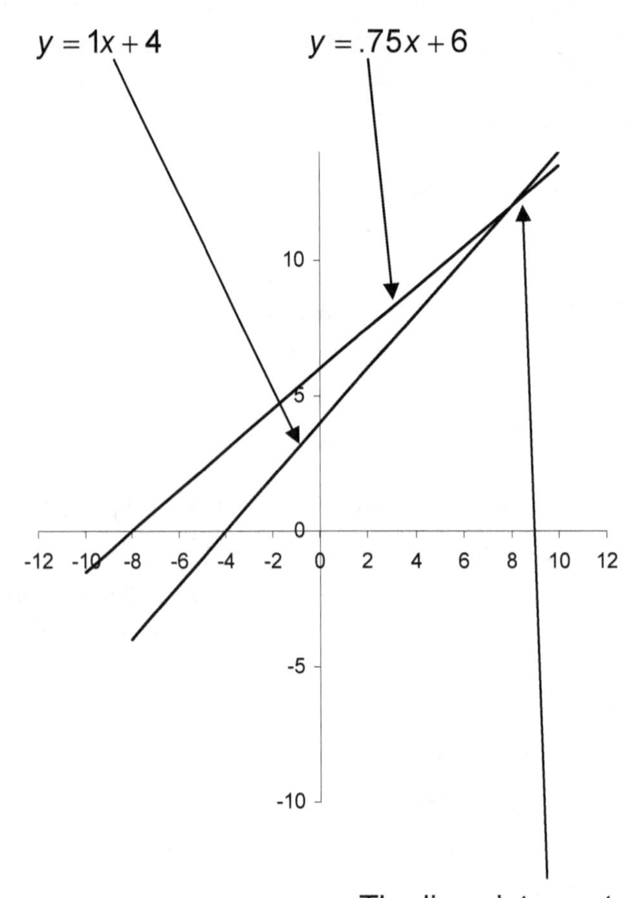

The lines intersect at (8, 12), therefore at 8 miles both companies charge $12. At values less than 8 miles, Aardvark Taxi charges less (the graph is below Baboon). When greater than 8 miles, Aardvark charges more (the graph is above Baboon).

If two things vary directly, as one gets larger, the other also gets larger. If one gets smaller, then the other gets smaller too. If x and y vary directly, there should be a constant, c, such that $y = cx$. Something can also vary directly with the square of something else, $y = cx^2$.

If two things vary inversely, as one gets larger, the other one gets smaller instead. If x and y vary inversely, there should be a constant, c, such that $xy = c$ or $y = c/x$. Something can also vary inversely with the square of something else, $y = c/x^2$.

Example: If $30 is paid for 5 hours work, how much would be paid for 19 hours work?

This is direct variation and $30 = 5c, so the constant is 6 ($6/hour). So $y = 6(19)$ or y = $114.

This could also be done as a proportion:

$$\frac{\$30}{5} = \frac{y}{19}$$
$$5y = 570$$
$$y = 114$$

Example: On a 546 mile trip from Miami to Charlotte, one car drove 65 mph while another car drove 70 mph. How does this affect the driving time for the trip?

This is an inverse variation, since increasing your speed should decrease your driving time. Using the equation: rate $\times$ time = distance, rt = d.

65t = 546	and	70t = 546
t = 8.4	and	t = 7.8

slower speed, more time   faster speed, less time

Example: A 14" pizza from Azzip Pizza costs $8.00. How much would a 20" pizza cost if its price was based on the same price per square inch? (font)

Here the price is directly proportional to the square of the radius. Using a proportion:

$$\frac{\$8.00}{7^2\pi} = \frac{x}{10^2\pi}$$
$$\frac{8}{153.86} = \frac{x}{314}$$
$$16.33 = x$$

$16.33 would be the price of the large pizza.

-The **absolute value function** for a 1st degree equation is of the form: $y = m(x - h) + k$. Its graph is in the shape of a $\vee$. The point (h,k) is the location of the maximum/minimum point on the graph. "$\pm$ m" are the slopes of the 2 sides of the $\vee$. The graph opens up if m is positive and down if m is negative.

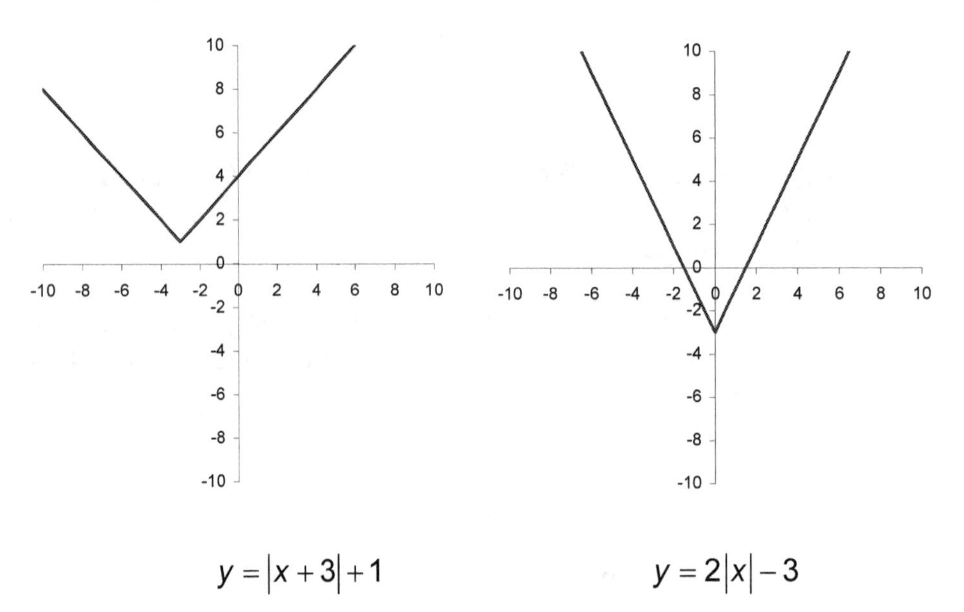

$y = |x + 3| + 1$ $\qquad$ $y = 2|x| - 3$

To solve an **absolute value equation**, follow these steps:

1. Get the absolute value expression alone on one side of the equation.

2. Split the absolute value equation into 2 separate equations without absolute value bars. Write the expression inside the absolute value bars (without the bars) equal to the expression on the other side of the equation. Now write the expression inside the absolute value bars equal to the opposite of the expression on the other side of the equation.

3. Now solve each of these equations.

4. **Check each answer by substituting it into the original equation** (with the absolute value symbol). There will be answers that do not check in the original equation. These answers are discarded as they are **extraneous solutions**. If all answers are discarded as incorrect, then the answer to the equation is $\varnothing$, which means the empty set or the null set. (0, 1, or 2 solutions could be correct.)

To solve an absolute value inequality, follow these steps:

1. Get the absolute value expression alone on one side of the inequality. Remember:

    Dividing or multiplying by a negative number will reverse the direction of the inequality sign.

2. Remember what the inequality sign is at this point.

3. Split the absolute value inequality into 2 separate inequalities without absolute value bars. First rewrite the inequality without the absolute bars and solve it. Next write the expression inside the absolute value bar followed by the opposite inequality sign and then by the opposite of the expression on the other side of the inequality. Now solve it.

4. If the sign in the inequality on step 2 is $<$ or $\leq$, the answer is those 2 inequalities connected by the word **and**. The solution set consists of the points between the 2 numbers on the number line. If the sign in the inequality on step 2 is $>$ or $\geq$, the answer is those 2 inequalities connected by the word **or**. The solution set consists of the points outside the 2 numbers on the number line.

If an expression inside an absolute value bar is compared to a negative number, the answer can also be either all real numbers or the empty set ($\varnothing$). For instance,

$$|x+3| < {}^{-}6$$

would have the empty set as the answer, since an absolute value is always positive and will never be less than ${}^{-}6$.

However,

$$|x+3| > {}^{-}6$$

would have all real numbers as the answer, since an absolute value is always positive or at least zero, and will never be less than -6. In similar fashion, (font)

$$|x+3| = {}^{-}6$$

would never check because an absolute value will never give a negative value.

<u>Example</u>: Solve and check:

$$|2x-5|+1=12$$

$|2x-5|=11$ Get absolute value alone.

Rewrite as 2 equations and solve separately.

same equation without absolute value		same equation without absolute value but right side is opposite
$2x-5=11$		$2x-5={}^{-}11$
$2x=16$	and	$2x={}^{-}6$
$x=8$		$x={}^{-}3$

Checks:

$\lvert 2x-5\rvert+1=12$	$\lvert 2x-5\rvert+1=12$
$\lvert 2(8)-5\rvert+1=12$	$\lvert 2(-3)-5\rvert+1=12$
$\lvert 11\rvert+1=12$	$\lvert -11\rvert+1=12$
$12=12$	$12=12$

This time both 8 and ${}^{-}3$ check.

Example: Solve and check:

$$2|x-7|-13 \geq 11$$

$$2|x-7| \geq 24$$ Get absolute value alone.

$$|x-7| \geq 12$$

Rewrite as 2 inequalities and solve separately.

same inequality without absolute value		same inequality without absolute value but right side and inequality sign are both the opposite
$x-7 \geq 12$	or	$x-7 \leq {}^{-}12$
$x \geq 19$	or	$x \leq {}^{-}5$

Competency 0011 Understand principles, concepts, and procedures related to measurement.

The strategy for solving problems of this nature should be to identify the given shapes and choose the correct formulas. Subtract the smaller cutout shape from the larger shape.
Sample problems:
Sample problems:

1. Find the area of one side of the metal in the circular flat washer shown below:

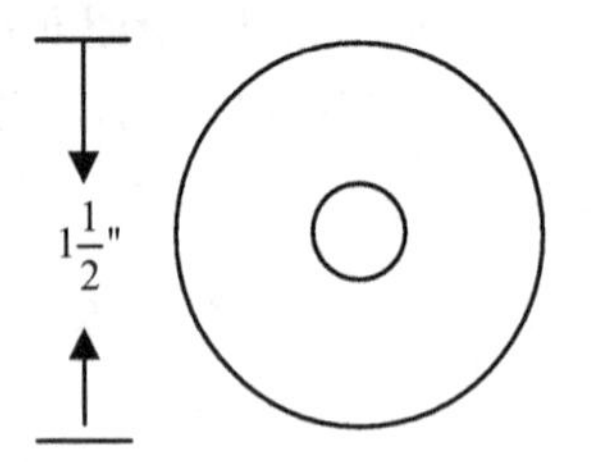

1. the shapes are both circles.
2. use the formula $A = \pi r^2$ for both.
   (**Inside diameter** is 3/8")

Area of larger circle	Area of smaller circle
$A = \pi r^2$	$A = \pi r^2$
$A = \pi(.75^2)$	$A = \pi(.1875^2)$
$A = 1.76625 \text{ in}^2$	$A = .1103906 \text{ in}^2$

Area of metal washer = larger area - smaller area

$= 1.76625 \text{ in}^2 - .1103906 \text{ in}^2$

$= 1.6558594 \text{ in}^2$

2. You have decided to fertilize your lawn. The shapes and dimensions of your lot, house, pool, and garden are given in the diagram below. The shaded area will not be fertilized. If each bag of fertilizer costs $7.95 and covers 4,500 square feet, find the total number of bags needed and the total cost of the fertilizer.

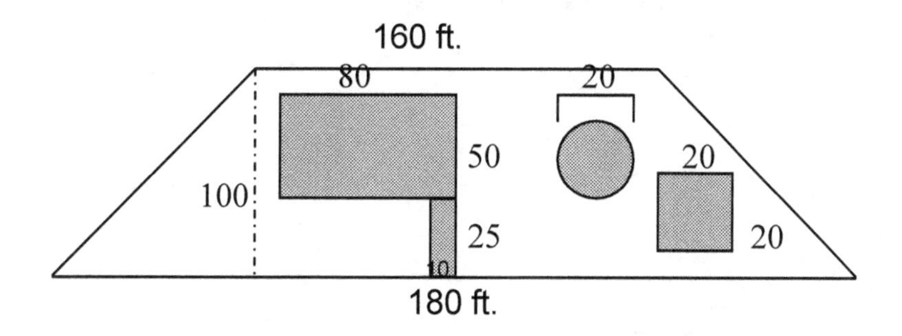

Area of Lot

$A = \frac{1}{2} h(b_1 + b_2)$

$A = \frac{1}{2}(100)(180 + 160)$

$A = 17{,}000$ sq ft

Area of House

$A = LW$

$A = (80)(50)$

$A = 4{,}000$ sq ft

Area of Driveway

$A = LW$

$A = (10)(25)$

$A = 250$ sq ft

Area of Pool

$A = \pi r^2$

$A = \pi(10)^2$

$A = 314.159$ sq. ft.

Area of Garden

$A = s^2$

$A = (20)^2$

$A = 400$ sq. ft.

Total area to fertilize = Lot area – (House + Driveway + Pool + Garden)

= 17,000 – (4,000 + 250 + 314.159 + 400)
= 12,035.841 sq ft

Number of bags needed = Total area to fertilize / 4,500 sq.ft. bag

= 12,035.841 / 4,500
= 2.67 bags

Since we cannot purchase 2.67 bags we must purchase 3 full bags.

Total cost = Number of bags * $7.95
= 3 * $7.95
= $23.85

Cut the **compound shape** into smaller, more familiar shapes and then compute the total area by adding the areas of the smaller parts.

Sample problem:

Find the area of the given shape.

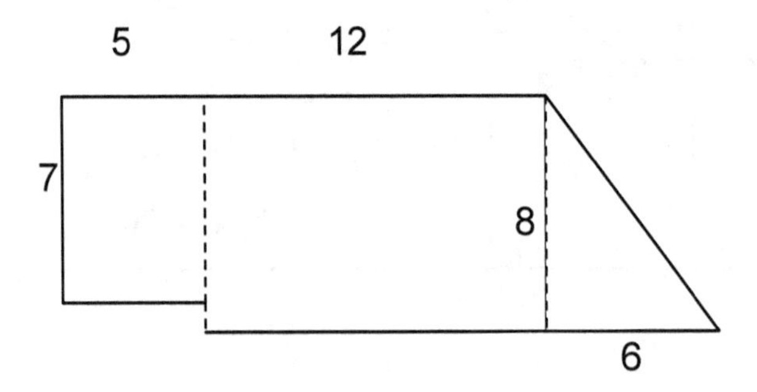

1. Using a dotted line we have cut the shape into smaller parts that are familiar.

2. Use the appropriate formula for each shape and find the sum of all areas.

Area 1 = LW	Area 2 = LW	Area 3 = ½bh
= (5)(7)	= (12)(8)	= ½(6)(8)
= 35 units^2	= 96 units^2	= 24 units^2

Total area = Area 1 + Area 2 + Area 3
= 35 + 96 + 24
= 155 units^2

It is necessary to be familiar with the metric and customary (English) system in order to estimate measurements.

Some common equivalents include:

ITEM	APPROXIMATELY EQUAL TO	
	METRIC	IMPERIAL
large paper clip	1 gram	1 ounce
1 quart	1 liter	
average sized man	75 kilograms	170 pounds
1 yard	1 meter	
math textbook	1 kilogram	2 pounds
1 mile	1 kilometer	
1 foot	30 centimeters	
thickness of a dime	1 millimeter	0.1 inches

Estimate the measurement of the following items:

The length of an adult cow = ________ meters
The thickness of a compact disc = ___________ millimeters
Your height = __________ meters
length of your nose = ___________ centimeters
weight of your math textbook = __________ kilograms
weight of an automobile = ___________ kilograms
weight of an aspirin = ___________ grams

Given a set of objects and their measurements, the use of rounding procedures is helpful when attempting to round to the nearest given unit.

When rounding to a given place value, it is necessary to look at the number in the next smaller place. If this number is 5 or more, the number in the place we are rounding to is increased by one; and all numbers to the right are changed to zero. If the number is less than 5, the number in the place we are rounding to stays the same; and all numbers to the right are changed to zero.

One method of rounding measurements can require an additional step. First, the measurement must be converted to a decimal number. Then the rules for rounding are applied.

Sample problem:

1. Round the measurements to the given units.

MEASUREMENT	ROUND TO NEAREST	ANSWER
1 foot 7 inches	foot	2 ft
5 pound 6 ounces	pound	5 pounds
5 9/16 inches	inch	6 inches

Solution:

Convert each measurement to a decimal number. Then apply the rules for rounding.

1 foot 7 inches = $1\frac{7}{12}$ ft = 1.58333 ft, round up to 2 ft

5 pounds 6 ounces = $5\frac{6}{16}$ pounds = 5.375 pound, round to 5 pounds

$5\frac{9}{16}$ inches = 5.5625 inches, round up to 6 inches

There are many methods for converting measurements within a system. One method is to multiply the given measurement by a conversion factor. This conversion factor is the ratio of:

$$\frac{\text{new units}}{\text{old units}} \quad \text{OR} \quad \frac{\text{what you want}}{\text{what you have}}$$

Sample problems:

1. Convert 3 miles to yards.

$$\frac{3 \text{ miles}}{1} \times \frac{1{,}760 \text{ yards}}{1 \text{ mile}} = \underline{\qquad\qquad} \text{yards}$$

=5,280 yards

1. Multiply by the conversion factor
2. Cancel the miles units
3. Solve

2. Convert 8,750 meters to kilometers.

$$\frac{8{,}750 \text{ meters}}{1} \times \frac{1 \text{ kilometer}}{1000 \text{ meters}} = \underline{\qquad\qquad} \text{km}$$

= 8.75 kilometers

1. Multiply by the conversion factor
2. Cancel the meters units
3. Solve

Most numbers in mathematics are "exact" or "counted". Measurements are "approximate". They usually involve interpolation or figuring out which mark on the ruler is closest. Any measurement you get with a measuring device is approximate. Variations in measurement are called precision and accuracy.

Precision is a measurement of how exactly a measurement is made, without reference to a true or real value. If a measurement is precise it can be made again and again with little variation in the result. The precision of a measuring device is the smallest fractional or decimal division on the instrument. The smaller the unit or fraction of a unit on the measuring device, the more precisely it can measure.

The greatest possible error of measurement is always equal to one-half the smallest fraction of a unit on the measuring device.

Accuracy is a measure of how close the result of measurement comes to the "true" value.

If you are throwing darts, the true value is the bull's eye. If the three darts land on the bull's eye, the dart thrower is both precise (all land near the same spot) and accurate (the darts all land on the "true" value).

The greatest measure of error allowed is called the tolerance. The least acceptable limit is called the lower limit and the greatest acceptable limit is called the upper limit. The difference between the upper and lower limits is called the tolerance interval. For example, a specification for an automobile part might be 14.625 ± 0.005 mm. This means that the smallest acceptable length of the part is 14.620 mm and the largest length acceptable is 14.630 mm. The tolerance interval is 0.010 mm. One can see how it would be important for automobile parts to be within a set of limits in terms of length. If the part is too long or too short it will not fit properly and vibrations will occur weakening the part and eventually causing damage to other parts.

Use the formulas to find the volume and surface area.

FIGURE	VOLUME	TOTAL SURFACE AREA
Right Cylinder	$\pi r^2 h$	$2\pi rh + 2\pi r^2$
Right Cone	$\frac{\pi r^2 h}{3}$	$\pi r\sqrt{r^2 + h^2} + \pi r^2$
Sphere	$\frac{4}{3}\pi r^3$	$4\pi r^2$
Rectangular Solid	LWH	$2LW + 2WH + 2LH$

Note: $\sqrt{r^2 + h^2}$ is equal to the slant height of the cone.

Sample problem:

1. Given the figure below, find the volume and surface area.

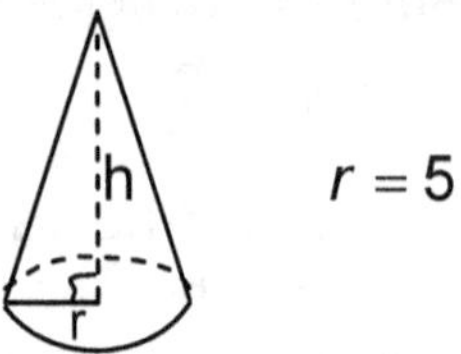

$r = 5$ in $\quad h = 6.2$ in

Volume $= \frac{\pi r^2 h}{3}$ First, write the formula.

$\frac{1}{3}\pi(5^2)(6.2)$ Then substitute.

162.3 cubic inches Finally, solve the problem.

Surface area = $\pi r\sqrt{r^2 + h^2} + \pi r^2$ First, write the formula.

$\pi 5\sqrt{5^2 + 6.2^2} + \pi 5^2$ Then substitute.

203.6 square inches Compute.

Note: volume is always given in cubic units and area is always given in square units.

FIGURE	AREA FORMULA	PERIMETER FORMULA
Rectangle	LW	$2(L+W)$
Triangle	$\frac{1}{2}bh$	$a+b+c$
Parallelogram	bh	sum of lengths of sides
Trapezoid	$\frac{1}{2}h(a+b)$	sum of lengths of sides

Sample problems:

1. Find the area and perimeter of a rectangle if its length is 12 inches and its diagonal is 15 inches.

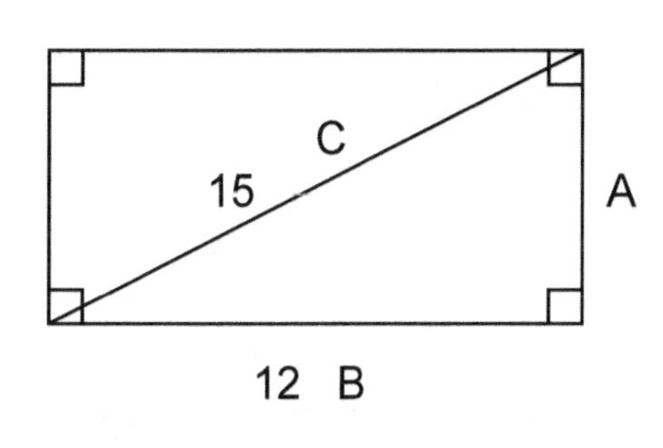

1. Draw and label sketch.

2. Since the height is still needed use Pythagorean formula to find missing leg of the triangle.

$$A^2 + B^2 = C^2$$
$$A^2 + 12^2 = 15^2$$
$$A^2 = 15^2 - 12^2$$
$$A^2 = 81$$
$$A = 9$$

Now use this information to find the area and perimeter.

$\text{A} = LW$	$\text{P} = 2(L+W)$	1. write formula
$\text{A} = (12)(9)$	$\text{P} = 2(12+9)$	2. substitute
$\text{A} = 108 \text{ in}^2$	$\text{P} = 42$ inches	3. solve

Given a circular figure the formulas are as follows:

$A = \pi r^2$ $\quad\quad$ $C = \pi d$ or $2\pi r$

Sample problem:

1. If the area of a circle is 50 cm^2, find the circumference.

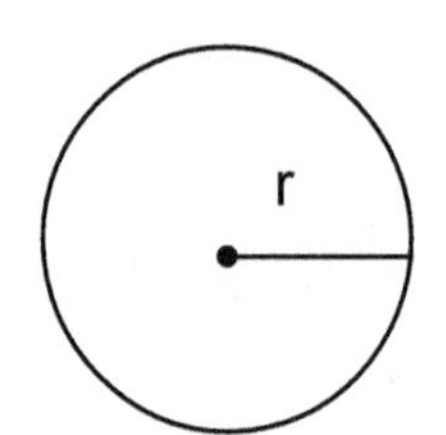

$A = 50 \text{ cm}^2$

1. Draw sketch.

2. Determine what is still needed.

Use the area formula to find the radius.

$A = \pi r^2$	1. Write formula
$50 = \pi r^2$	2. Substitute
$\frac{50}{\pi} = r^2$	3. Divide by π
$15.915 = r^2$	4. Substitute
$\sqrt{15.915} = \sqrt{r^2}$	5. Take square root of both sides
$3.989 \approx r$	6. Compute

Use the approximate answer (due to rounding) to find the circumference.

$C = 2\pi r$	1. Write formula
$C = 2\pi\,(3.989)$	2. Substitute
$C \approx 25.064$	3. Compute

Competency 0012 Understand the principles of Euclidean geometry and use them to prove theorems.

In geometry the point, line and plane are key concepts and can be discussed in relation to each other.

collinear points
are all on the same line

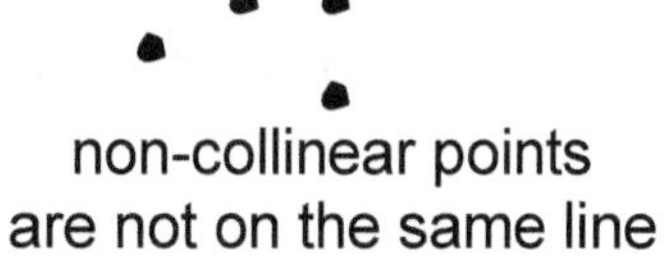

non-collinear points
are not on the same line

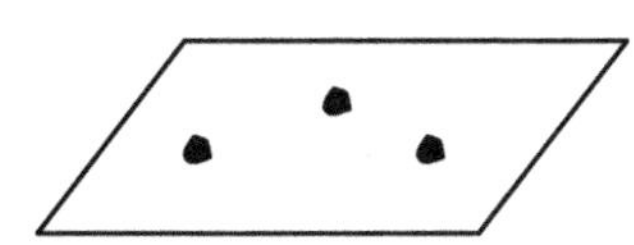

coplanar points
are on the same plane

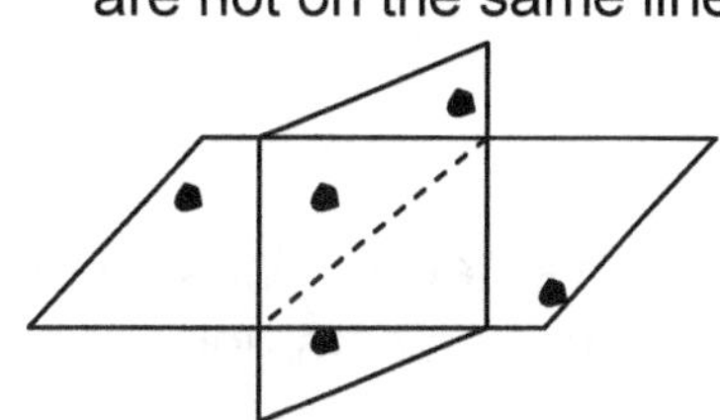

non-coplanar points
are not on the same plane

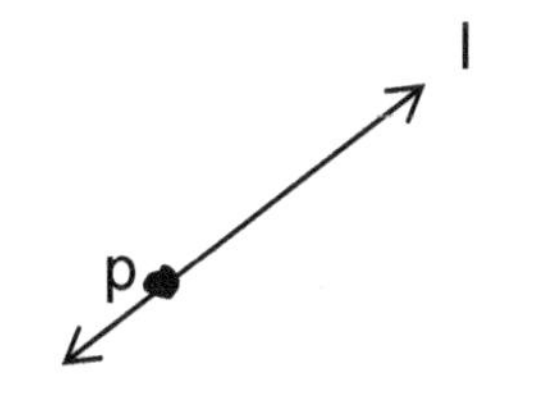

Point p is on line l
l contains P
l passes through P

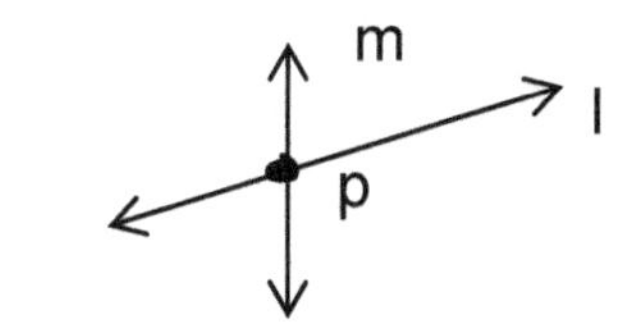

l and m intersect
at p
p is the intersection
of l and m

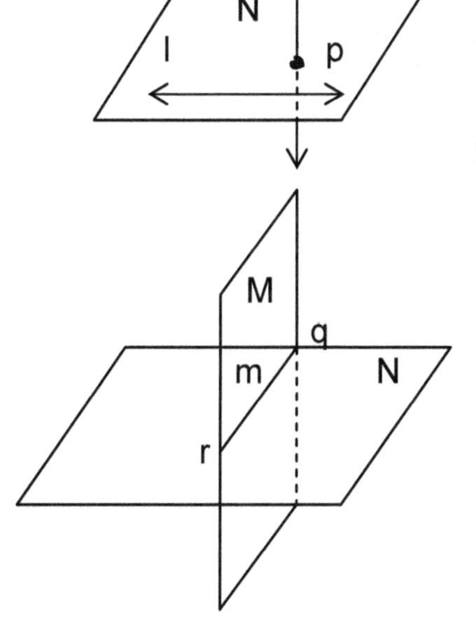

l and p are in plane N
N contains p and l
Line m intersects N at p
p is the intersection
of m and N

Planes M and N intersect at rq
rq is the intersection
of M and N
rq is in M and N
M and N contain rq

Parallel lines or planes do not intersect.

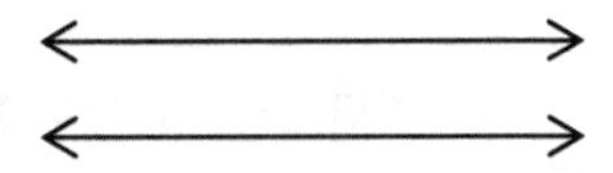

Perpendicular lines or planes form a 90 degree angle to each other.

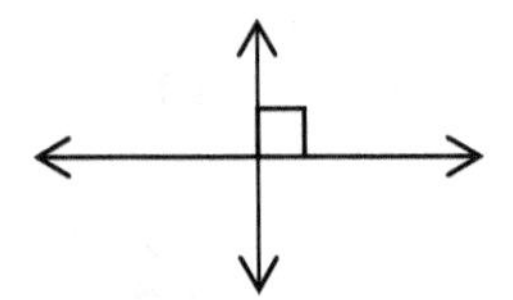

Intersecting lines share a common point and intersecting planes share a common set of points or line.

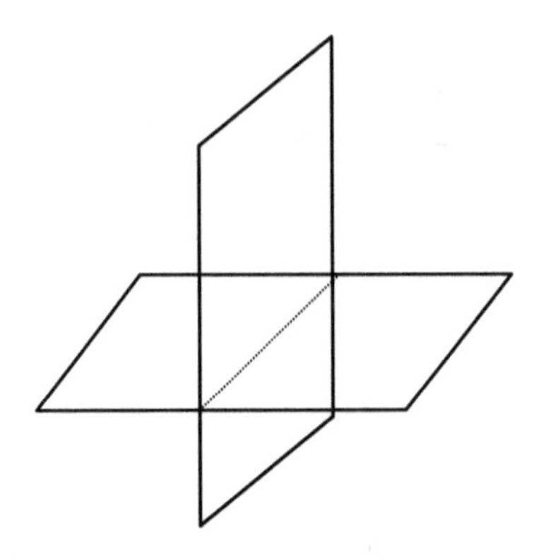

Skew lines do not intersect and do not lie on the same plane.

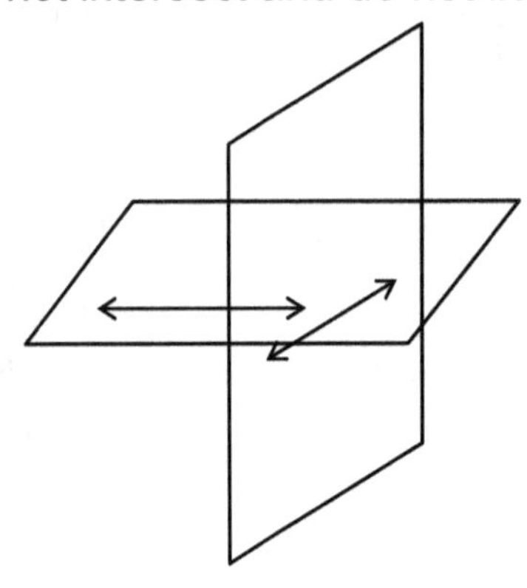

Congruent figures have the same size and shape. If one is placed above the other, it will fit exactly. Congruent lines have the same length. Congruent angles have equal measures.
The symbol for congruent is $\cong$.

Polygons (pentagons) *ABCDE* and *VWXYZ* are congruent. They are exactly the same size and shape.

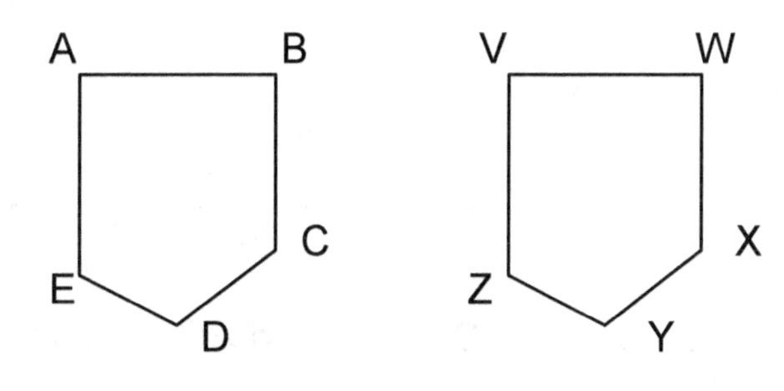

$ABCDE \cong VWXYZ$

Corresponding parts are those congruent angles and congruent sides, that is:

corresponding angles	corresponding sides
$\angle A \leftrightarrow \angle V$	$AB \leftrightarrow VW$
$\angle B \leftrightarrow \angle W$	$BC \leftrightarrow WX$
$\angle C \leftrightarrow \angle X$	$CD \leftrightarrow XY$
$\angle D \leftrightarrow \angle Y$	$DE \leftrightarrow YZ$
$\angle E \leftrightarrow \angle Z$	$AE \leftrightarrow VZ$

Two triangles can be proven congruent by comparing pairs of appropriate congruent corresponding parts.

SSS POSTULATE

If three sides of one triangle are congruent to three sides of another triangle, then the two triangles are congruent.

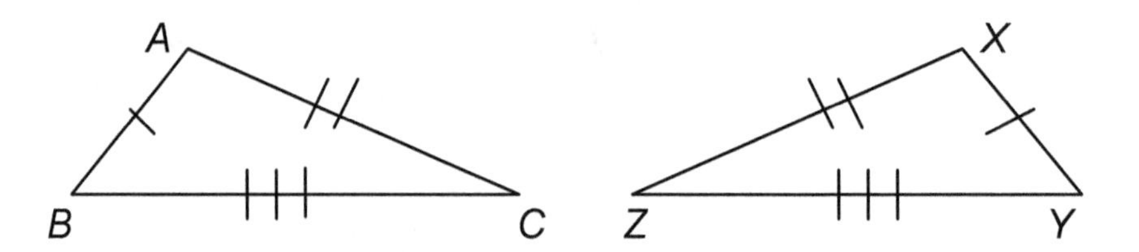

Since $AB \cong XY$, $BC \cong YZ$ and $AC \cong XZ$, then $\Delta ABC \cong \Delta XYZ$.

Example: Given isosceles triangle *ABC* with *D* the midpoint of base *AC*, prove the two triangles formed by *AD* are congruent.

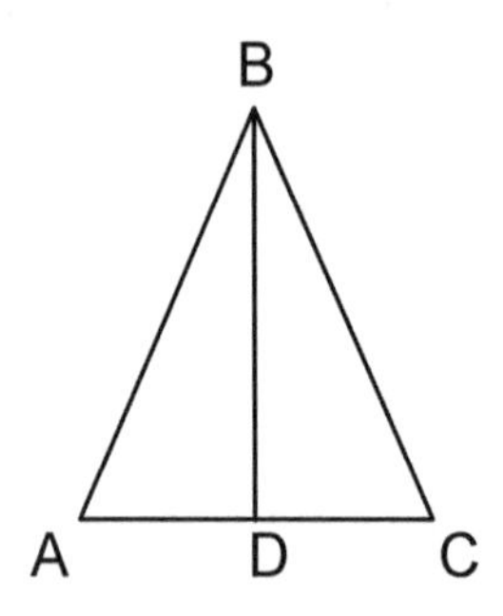

Proof:

1. Isosceles triangle *ABC*, *D* midpoint of base *AC*	Given
2. $AB \cong BC$	An isosceles Δ has two congruent sides
3. $AD \cong DC$	Midpoint divides a line into two equal parts
4. $BD \cong BD$	Reflexive
5. $\Delta ABD \cong \Delta BCD$	SSS

SAS POSTULATE

If two sides and the included angle of one triangle are congruent to two sides and the included angle of another triangle, then the two triangles are congruent.

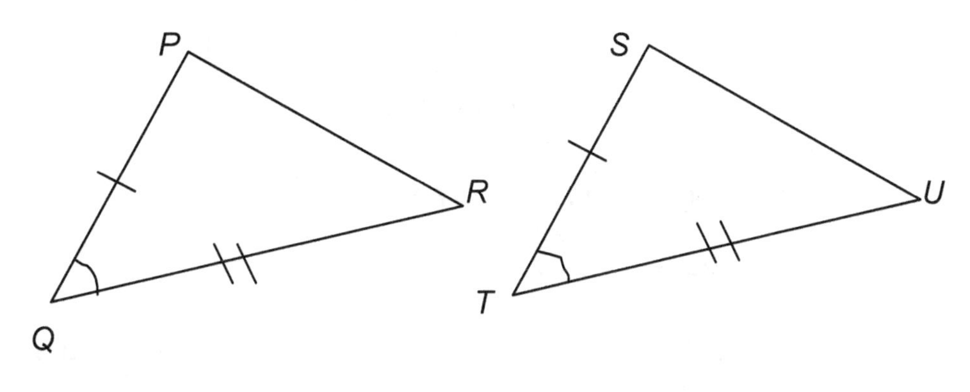

Example:

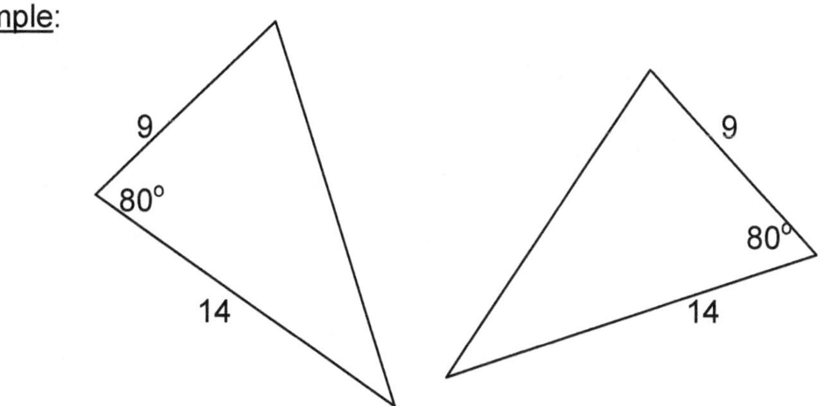

The two triangles are congruent by SAS.

ASA POSTULATE

If two angles and the included side of one triangle are congruent to two angles and the included side of another triangle, the triangles are congruent.

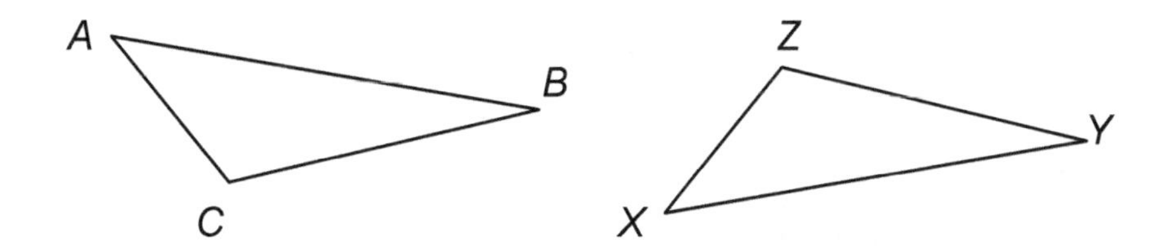

$\angle A \cong \angle X$, $\angle B \cong \angle Y$, $AB \cong XY$ then $\Delta ABC \cong \Delta XYZ$ by ASA

Example: Given two right triangles with one leg of each measuring 6 cm and the adjacent angle 37°, prove the triangles are congruent.

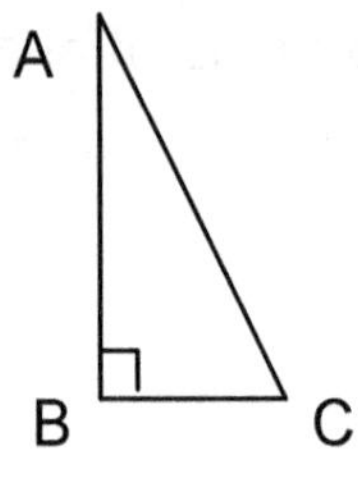

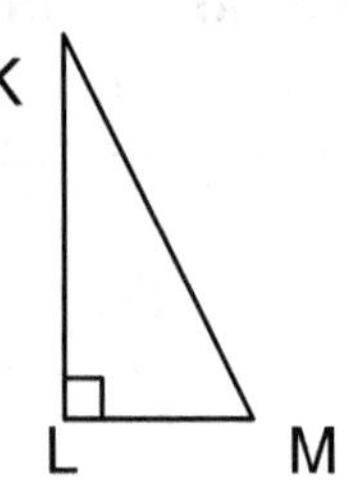

1. Right triangles ABC and KLM $AB = KL = 6$ cm $\angle A = \angle K = 37^{\circ}$	Given
2. $AB \cong KL$ $\angle A \cong \angle K$	Figures with the same measure are congruent
3. $\angle B \cong \angle L$	All right angles are congruent.
4. $\Delta ABC \cong \Delta KLM$	ASA

Example:
What method would you use to prove the triangles congruent?

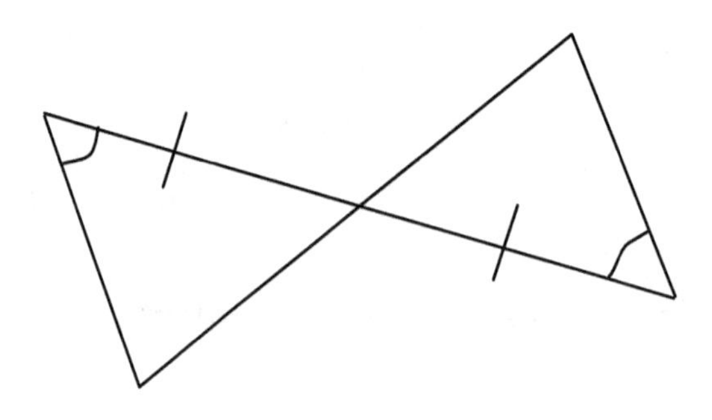

ASA because vertical angles are congruent.

AAS THEOREM

If two angles and a non-included side of one triangle are congruent to the corresponding parts of another triangle, then the triangles are congruent.

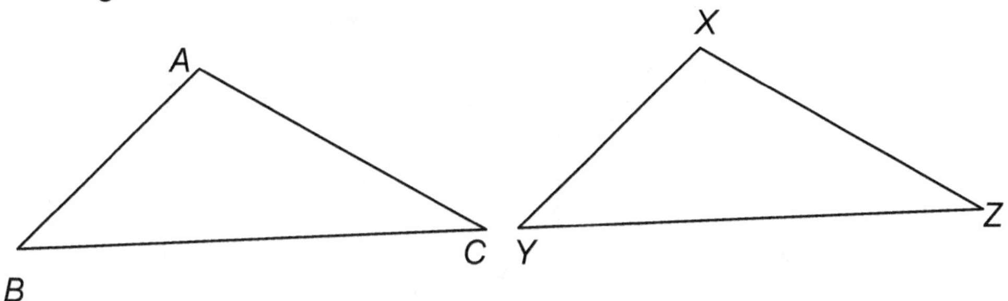

$\angle B \cong \angle Y$, $\angle C \cong \angle Z$, $AC \cong XZ$, then $\Delta ABC \cong \Delta XYZ$ by AAS.
We can derive this theorem because if two angles of the triangles are congruent, then the third angle must also be congruent. Therefore, we can use the ASA postulate.

HL THEOREM

If the hypotenuse and a leg of one right triangle are congruent to the corresponding parts of another right triangle, the triangles are congruent.

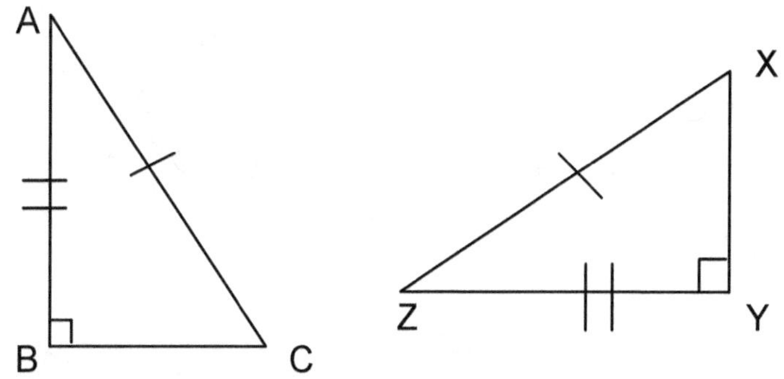

Since $\angle B$ and $\angle Y$ are right angles and $AC \cong XZ$ (hypotenuse of each triangle), $AB \cong YZ$ (corresponding leg of each triangle), then $\Delta ABC \cong \Delta XYZ$ by HL.

Example: What method would you use to prove the triangles congruent?

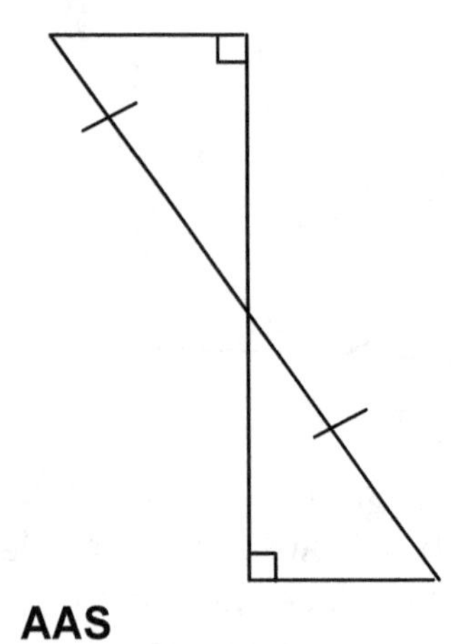

HL

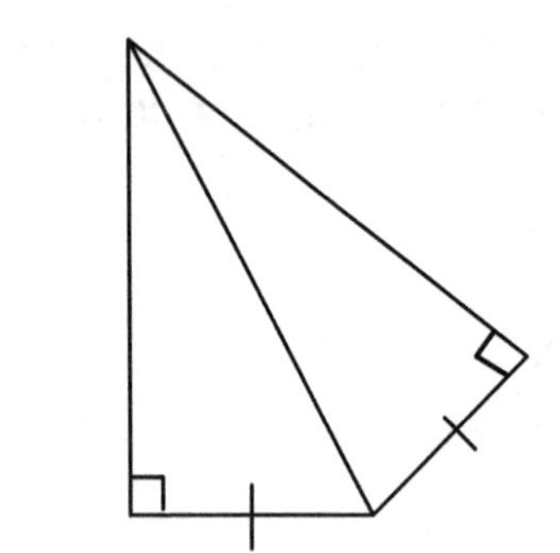

The 3 undefined terms of geometry are point, line, and plane.

A plane is a flat surface that extends forever in two dimensions. It has no ends or edges. It has no thickness. It is usually drawn as a parallelogram that can be named either by 3 non-collinear points (3 points that are not on the same line) on the plane or by placing a letter in the corner of the plane that is not used elsewhere in the diagram.

A line extends forever in one dimension. It is determined and named by 2 points that are on the line. The line consists of every point that is between those 2 points as well as the points that are on the "straight" extension each way. A line is drawn as a line segment with arrows facing opposite directions on each end to indicate that the line continues in both directions forever.

A point is a position in space, on a line, or on a plane. It has no thickness and no width. Only 1 line can go through any 2 points. A point is represented by a dot named by a single letter.

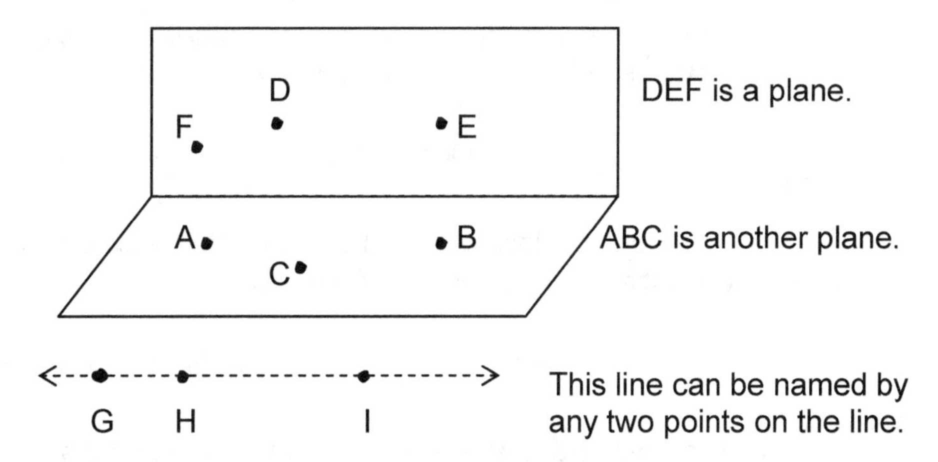

It could be named $\overleftrightarrow{GH}$, $\overleftrightarrow{HI}$, $\overleftrightarrow{GI}$, $\overleftrightarrow{IG}$, $\overleftrightarrow{IH}$, or $\overleftrightarrow{HG}$. Any 2 points (letters) on the line can be used and their order is not important in naming a line.

In the above diagrams, A, B, C, D, E, F, G, H, and I are all locations of individual points.

A ray is not an undefined term. A ray consists of all the points on a line starting at one given point and extending in only one of the two opposite directions along the line. The ray is named by naming 2 points on the ray. The first point must be the endpoint of the ray, while the second point can be any other point along the ray. The symbol for a ray is a ray above the 2 letters used to name it. The endpoint of the ray MUST be the first letter.

This ray could be named $\overrightarrow{JK}$ or $\overrightarrow{JL}$. It cannot be called $\overrightarrow{KJ}$ or $\overrightarrow{LJ}$ or $\overrightarrow{LK}$ or $\overrightarrow{KL}$ because none of those names start with the endpoint, J.

The **distance** between 2 points on a number line is equal to the absolute value of the difference of the two numbers associated with the points.

If one point is located at "a" and the other point is at "b", then the distance between them is found by this formula:

$$\text{distance} = |a - b| \text{ or } |b - a|$$

If one point is located at $^{-}3$ and another point is located at 5, the distance between them is found by:

$$\text{distance} = |a - b| = |(^{-}3) - 5| = |^{-}8| = 8$$

The only **undefined terms** are point, line and plane.

Definitions are explanations of all mathematical terms except those that are undefined.

Postulates are mathematical statements that are accepted as true statements without providing a proof.

Theorems are mathematical statements that can be proven to be true based on postulates, definitions, algebraic properties, given information, and previously proved theorems.

The following algebraic postulates are frequently used as reasons for statements in 2 column geometric properties:

Addition Property:

If $a = b$ and $c = d$, then $a + c = b + d$.

Subtraction Property:

If $a = b$ and $c = d$, then $a - c = b - d$.

Multiplication Property:

If $a = b$ and $c \neq 0$, then $ac = bc$.

Division Property:

If $a = b$ and $c \neq 0$, then $a/c = b/c$.

Reflexive Property:	$a = a$
Symmetric Property:	If $a = b$, then $b = a$.
Transitive Property:	If $a = b$ and $b = c$, then $a = c$.
Distributive Property:	$a(b + c) = ab + ac$
Substitution Property:	If $a = b$, then b may be substituted for a in any other expression (a may also be substituted for b).

In a **2 column proof**, the left side of the proof should be the given information, or statements that could be proved by deductive reasoning. The right column of the proof consists of the reasons used to determine that each statement to the left was verifiably true. The right side can identify given information, or state theorems, postulates, definitions or algebraic properties used to prove that particular line of the proof is true.

Indirect Proofs

Assume the opposite of the conclusion. Keep your hypothesis and given information the same. Proceed to develop the steps of the proof, looking for a statement that contradicts your original assumption or some other known fact. This contradiction indicates that the assumption you made at the beginning of the proof was incorrect; therefore, the original conclusion has to be true.

Inductive thinking is the process of finding a pattern from a group of examples. That pattern is the conclusion that this set of examples seemed to indicate. It may be a correct conclusion or it may be an incorrect conclusion because other examples may not follow the predicted pattern.

Deductive thinking is the process of arriving at a conclusion based on other statements that are all known to be true, such as theorems, axiomspostulates, or postulates. Conclusions found by deductive thinking based on true statements will **always** be true. Examples :

Suppose:
On Monday Mr.Peterson eats breakfast at McDonalds.
On Tuesday Mr.Peterson eats breakfast at McDonalds.
On Wednesday Mr.Peterson eats breakfast at McDonalds.
On Thursday Mr.Peterson eats breakfast at McDonalds again.

Conclusion: On Friday Mr. Peterson will eat breakfast at McDonalds again.

This is a conclusion based on inductive reasoning. Based on several days' observations, you conclude that Mr. Peterson will eat at McDonalds. This may or may not be true, but it is a conclusion arrived at by inductive thinking.

Two triangles are overlapping if a portion of the interior region of one triangle is shared in common with all or a part of the interior region of the second triangle.

The most effective method for proving two overlapping triangles congruent is to draw the two triangles separately. Separate the two triangles and label all of the vertices using the labels from the original overlapping figures. Once the separation is complete, apply one of the congruence shortcuts: SSS, ASA, SAS, AAS, or HL.

A trapezoid is a quadrilateral with exactly one pair of parallel sides. A trapezoid is different from a parallelogram because a parallelogram has two pairs of parallel sides.

The two parallel sides of a trapezoid are called the bases, and the two non-parallel sides are called the legs. If the two legs are the same length, then the trapezoid is called isosceles.

The segment connecting the two midpoints of the legs is called the **median**. The median has the following two properties.

The median is parallel to the two bases.

The length of the median is equal to one-half the sum of the length of the two bases.

The segment joining the midpoints of two sides of a triangle is also called a **median**. All triangles have three medians. Each median has the following two properties.

A median is parallel to the third side of the triangle.

The length of a median is one-half the length of the third side of the triangle.

Every angle has exactly one ray which bisects the angle. If a point on such a bisector is located, then the point is equidistant from the two sides of the angle. The distance from a point to a side is measured along a segment which is perpendicular to the angle's side. The converse is also true. If a point is equidistant from the sides of an angle, then the point is on the bisector of the angle.

Every segment has exactly one line which is both perpendicular to and bisects the segment. If a point on such a perpendicular bisector is located, then the point is equidistant to the endpoints of the segment. The converse is also true. If a point is equidistant from the endpoints of a segment, then that point is on the perpendicular bisector of the segment.

If the three segments which bisect the three angles of a triangle are drawn, the segments will all intersect in a single point. This point is equidistant from all three sides of the triangle. Recall that the distance from a point to a side is measured along the perpendicular from the point to the side.

An **altitude** is a segment which extends from one vertex and is perpendicular to the side opposite that vertex. In some cases, the side opposite from the vertex used will need to be extended in order for the altitude to form a perpendicular to the opposite side. The length of the altitude is used when referring to the height of the triangle.

If two planes are parallel and a third plane intersects the first two, then the three planes will intersect in two lines which are also parallel.

Given both a line and a point which is not on the line but is in the same plane, then there is exactly one line through the point which is parallel to the given line and exactly one line through the point which is perpendicular to the given line.

If three or more segments intersect in a single point, the point is called a **point of concurrency**.

The following sets of special segments all intersect in points of concurrency.

1. Angle Bisectors
2. Medians
3. Altitudes
4. Perpendicular Bisectors

The points of concurrency can lie inside the triangle, outside the triangle, or on one of the sides of the triangle. The following table summarizes this information.

Possible Location(s) of the Points of Concurrency

	Inside the Triangle	Outside the Triangle	On the Triangle
Angle Bisectors	x		
Medians	x		
Altitudes	x	x	x
Perpendicular Bisectors	x	x	x

The point of concurrency of the three medians in a triangle is called the **centroid**.

The centroid divides each median into two segments whose lengths are always in the ratio of 1:2. The distance from the vertex to the centroid is always twice the distance from the centroid to the midpoint of the side opposite the vertex.

If **two circles** have radii which are in a ratio of $a:b$, then the following ratios are also true for the circles.

The diameters are also in the ratio $a:b$.
The circumferences are also in the ratio $a:b$.
The areas are in the ratio $a^2:b^2$, or the ratio of the areas is the square of the ratios of the radii.

Euclid wrote a set of 13 books around 330 B.C. called The Elements. He outlined ten axioms and then deduced 465 theorems. Euclidean geometry is based on the undefined concept of the point, line and plane.

The fifth of Euclid's axioms (referred to as the parallel postulate) was not as readily accepted as the other nine axioms. Many mathematicians throughout the years have attempted to prove that this axiom is not necessary because it could be proved by the other nine. Among the many who attempted to prove this was Carl Friedrich Gauss. His works led to the development of hyperbolic geometry. Elliptical or Reimannian geometry was suggested by G.F. Bernhard Riemann. He based his work on the theory of surfaces and used models as physical interpretations of the undefined terms that satisfy the axioms.

The chart below lists the fifth axiom (parallel postulate) as it is given in each of the three geometries.

EUCLIDEAN	ELLIPTICAL	HYPERBOLIC
Given a line and a point not on that line, one and only one line can be drawn through the given point parallel to the given line.	Given a line and a point not on that line, no line can be drawn through the given point parallel to the given line.	Given a line and a point not on that line, two or more lines can be drawn through the point parallel to the given line.

Origins and Development of Geometry

The field of *geometry* began as a crude collection of rules for calculating lengths, areas, and volumes. Ancient practitioners of construction, surveying, and navigation utilized basic geometric principles in performing their trades. Historians believe the Egyptians and Babylonians used geometry as early as 3000 B.C. Furthermore, based on the level of development of Chinese and Hindu culture during the same period, it is likely the Chinese and Hindu people used geometry.

The Greeks took ancient geometric thought and developed it into a true mathematical science. The evolution of Greek geometry began with Thales, who introduced the concept of theorems by proof, the defining characteristic of Greek mathematics. Next, Pythagoras and his disciples created a large body of geometric thought through deduction. Finally, Euclid organized and added to previous knowledge, by writing *Elements*, his thirteen-volume compilation of Greek geometry and number theory in 300 B.C. *Elements* was the first formal system of mathematics in which assumptions, called axioms and postulates, logically led to the formation of various theorems, corollaries, and definitions. Euclidean geometry, the system of geometry based on Euclid's *Elements*, is the traditional form of geometry studied in elementary and secondary schools today.

Though Greek geometry began to decline around 200 B.C., geometry continued to develop in other cultures. The Islamic empire contributed greatly to the development of geometry from 700 to 1500 A.D. For example, the Persian poet and mathematician Omar Khayyam created the field of algebraic geometry in the 11th Century and paved the way for the development of non-Euclidean geometry.

In the 17th Century, Rene Descartes and Pierre de Fermat created analytic geometry, the geometry of coordinates and equations. Analytic geometry is the precursor of calculus. In addition, Girard Desargues created projective geometry, the study of geometry without measurement.

The 18th and 19th centuries saw the development of non-Euclidean geometry or hyperbolic, elliptical and absolute geometry. Non-Euclidean geometry rejects Euclid's fifth postulate of parallel lines. Finally, the advances in technology in the 20th Century sparked the development of computational or digital geometry.

Early Development in Different Cultures

Geometry originated and began to develop in areas distant from Greece (e.g. India and China) as early as 3000 B.C. The early Hindu cultures of India and Pakistan extensively used weights and measures in construction, circles and triangles in art, and rudimentary compasses for navigation. Later, Hindu cultures formalized geometric concepts with rules and theorems. For example, the *Suba Sutras,* written in India between 800 and 500 B.C., contains geometric proofs and the Pythagorean Theorem.

Early Chinese culture also used geometry in construction and measurement. Among the geometric principles used by the early Chinese cultures were area formulas for two-dimensional figures, proportionality constants like π and the Pythagorean Theorem.

Competency 0013 Apply Euclidean geometry to analyze the properties of two-dimensional figures and to solve problems.

The classifying of angles refers to the angle measure. The naming of angles refers to the letters or numbers used to label the angle.

Sample Problem:

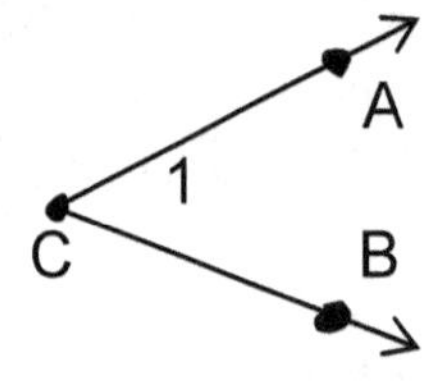

$\overrightarrow{CA}$ (read ray CA) and $\overrightarrow{CB}$ are the sides of the angle.
The angle can be called $\angle ACB$, $\angle BCA$, $\angle C$ or $\angle 1$.

Angles are classified according to their size as follows:

acute: greater than 0 and less than 90 degrees.
right: exactly 90 degrees.
obtuse: greater than 90 and less than 180 degrees.
straight: exactly 180 degrees

Angles can be classified in a number of ways. Some of those classifications are outlined here.

Adjacent angles have a common vertex and one common side but no interior points in common.

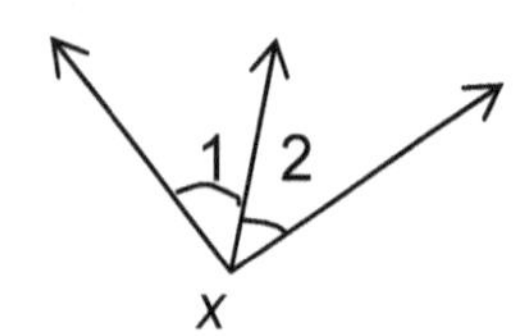

Complimentary angles add up to 90 degrees.

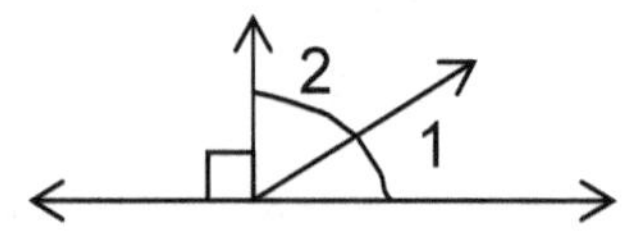

Supplementary angles add up to 180 degrees.

Vertical angles have sides that form two pairs of opposite rays.

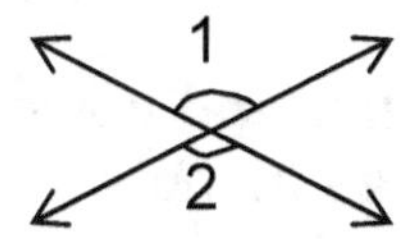

Corresponding angles are in the same corresponding position on two parallel lines cut by a transversal.

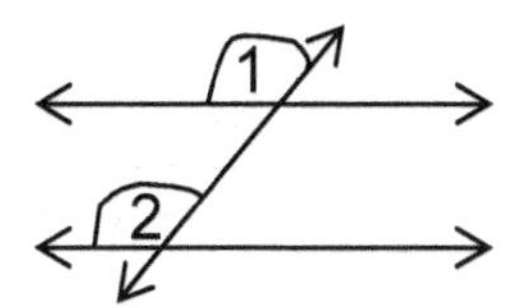

Alternate interior angles are diagonal angles on the inside of two parallel lines cut by a transversal.

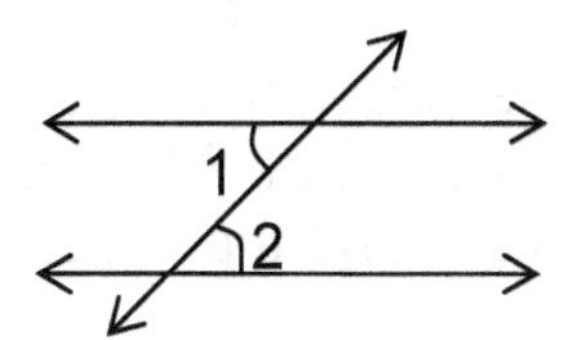

Alternate exterior angles are diagonal angles on the outside of two parallel lines cut by a transversal.

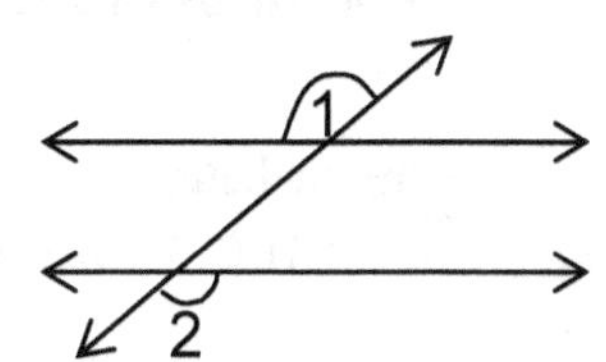

A **triangle** is a polygon with three sides.

Triangles can be classified by the types of angles or the lengths of their sides.

Classifying by angles:

An **acute** triangle has exactly three *acute* angles.
A **right** triangle has one *right* angle.
An **obtuse** triangle has one *obtuse* angle.

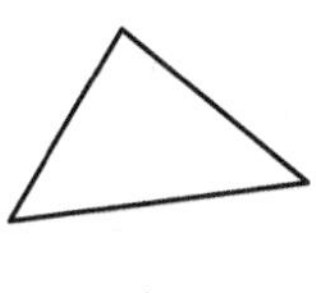

acute

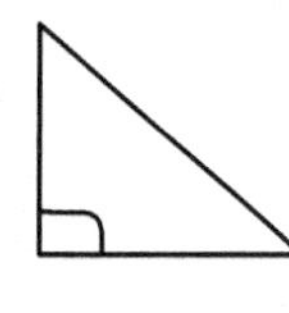

right

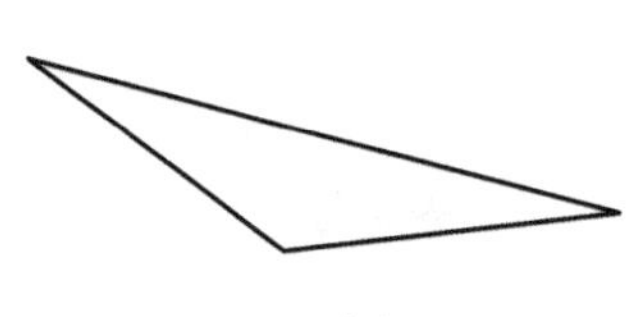

obtuse

Classifying by sides:

All *three* sides of an **equilateral** triangle are the same length.
Two sides of an **isosceles** triangle are the same length.
None of the sides of a **scalene** triangle are the same length.

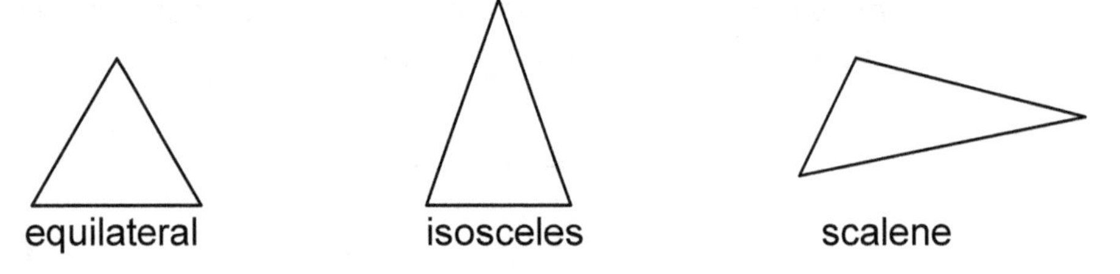

equilateral isosceles scalene

The sum of the measures of the angles of a triangle is 180°.

Example 1:
Can a triangle have two right angles?
No. A right angle measures 90°, therefore the sum of two right angles would be 180° and there could not be third angle.

Example 2:
Can a triangle have two obtuse angles?
No. Since an obtuse angle measures more than 90° the sum of two obtuse angles would be greater than 180°.

Example 3:
Can a right triangle be obtuse?
No. Once again, the sum of the angles would be more than 180°.

Example 4:
In a triangle, the measure of the second angle is three times the first. The third angle equals the sum of the measures of the first two angles. Find the number of degrees in each angle.

Let x = the number of degrees in the first angle
$3x$ = the number of degrees in the second angle
$x + 3x$ = the measure of the third angle

Since the sum of the measures of all three angles is 180°.

$$x + 3x + (x + 3x) = 180$$
$$8x = 180$$
$$x = 22.5$$
$$3x = 67.5$$
$$x + 3x = 90$$

Thus the angles measure 22.5°, 67.5°, and 90°. Additionally, the triangle is a right triangle.

EXTERIOR ANGLES

Two adjacent angles form a linear pair when they have a common side and their remaining sides form a straight angle. Angles in a linear pair are supplementary. An exterior angle of a triangle forms a linear pair with an angle of the triangle.

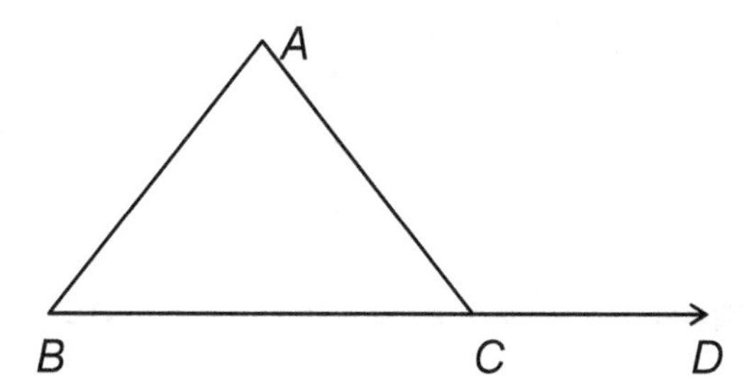

$\angle ACD$ is an exterior angle of triangle *ABC*, forming a linear pair with $\angle ACB$.

The measure of an exterior angle of a triangle is equal to the sum of the measures of the two non-adjacent interior angles.

Example:
In triangle *ABC*, the measure of $\angle A$ is twice the measure of $\angle B$. $\angle C$ is 30° more than their sum. Find the measure of the exterior angle formed at $\angle C$.

Let x = the measure of $\angle B$
$2x$ = the measure of $\angle A$
$x + 2x + 30$ = the measure of $\angle C$

$$x + 2x + x + 2x + 30 = 180$$
$$6x + 30 = 180$$
$$6x = 150$$
$$x = 25$$
$$2x = 50$$

It is not necessary to find the measure of the third angle, since the exterior angle equals the sum of the opposite interior angles. Thus the exterior angle at $\angle C$ measures 75°.

A **quadrilateral** is a polygon with four sides.
The sum of the measures of the angles of a quadrilateral is 360°.

A **trapezoid** is a quadrilateral with exactly <u>one</u> pair of parallel sides.

In an **isosceles trapezoid**, the non-parallel sides are congruent.

A **parallelogram** is a quadrilateral with <u>two</u> pairs of parallel sides.

A **rectangle** is a parallelogram with a right angle.

A **rhombus** is a parallelogram with all sides equal length.

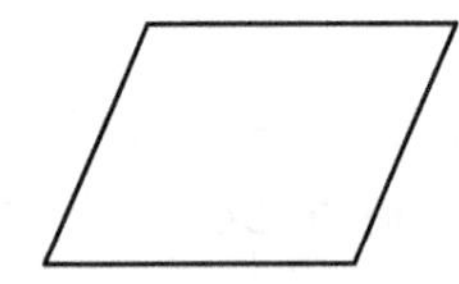

A **square** is a rectangle with all sides equal length.

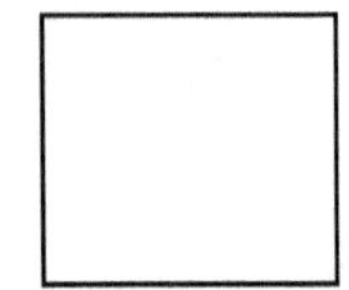

A **parallelogram** exhibits these properties.

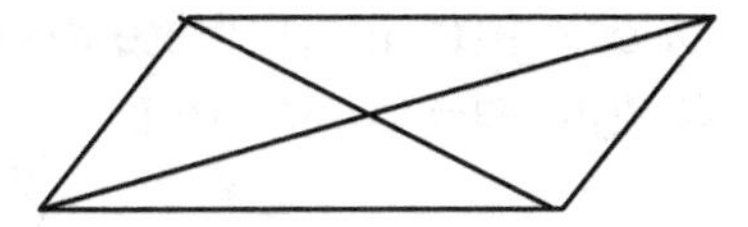

The diagonals bisect each other.

Each diagonal divides the parallelogram into two congruent triangles.

Both pairs of opposite sides are congruent.
Both pairs of opposite angles are congruent.
Two adjacent angles are supplementary.

Example 1:
Find the measures of the other three angles of a parallelogram if one angle measures 38°.

Since opposite angles are equal, there are two angles measuring 38°.

Since adjacent angles are supplementary, 180 - 38 = 142 so the other two angles measure 142 ° each.

$$\begin{array}{r} 38 \\ 38 \\ 142 \\ +\ \underline{142} \\ 360 \end{array}$$

Example 2:
The measures of two adjacent angles of a parallelogram are $3x + 40$ and $x + 70$.

Find the measures of each angle.

$$\begin{aligned} 2(3x + 40) + 2(x + 70) &= 360 \\ 6x + 80 + 2x + 140 &= 360 \\ 8x + 220 &= 360 \\ 8x &= 140 \\ x &= 17.5 \\ 3x + 40 &= 92.5 \\ x + 70 &= 87.5 \end{aligned}$$

Thus the angles measure 92.5°, 92.5°, 87.5°, and 87.5°.

Since a **rectangle** is a special type of parallelogram, it exhibits all the properties of a parallelogram. All the angles of a rectangle are right angles because of congruent opposite angles. Additionally, the diagonals of a rectangle are congruent.

A **rhombus** also has all the properties of a parallelogram. Additionally, its diagonals are perpendicular to each other and they bisect its angles.

A **square** has all the properties of a rectangle and a rhombus.

Example 1:

	True or false?
All squares are rhombuses.	True
All parallelograms are rectangles.	False - some parallelograms are rectangles
All rectangles are parallelograms.	True
Some rhombuses are squares.	True
Some rectangles are trapezoids.	False - only one pair of parallel sides
All quadrilaterals are parallelograms.	False -some quadrilaterals are parallelograms
Some squares are rectangles.	False - all squares are rectangles
Some parallelograms are rhombuses.	True

Example 2:

In rhombus $ABCD$ side $AB = 3x - 7$ and side $CD = x + 15$. Find the length of each side.

Since all the sides are the same length, $3x - 7 = x + 15$

$$2x = 22$$

$$x = 11$$

Since 3(11) - 7 = 25 and 11 + 15 = 25, each side measures 26 units.

A **trapezoid** is a quadrilateral with exactly one pair of parallel sides.

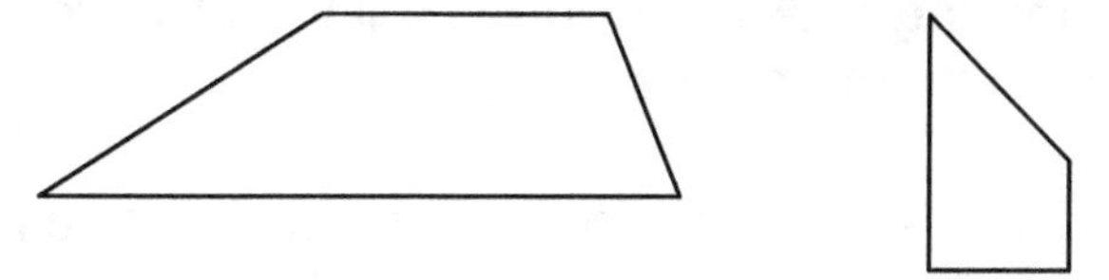

In an **isosceles trapezoid**, the non-parallel sides are congruent.

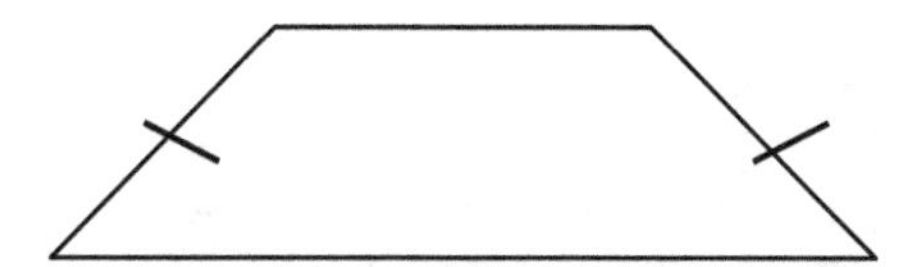

An isosceles trapezoid has the following properties:

The diagonals of an isosceles trapezoid are congruent.
The base angles of an isosceles trapezoid are congruent.

Example:

An isosceles trapezoid has a diagonal of 10 and a base angle measure of 30°. Find the measure of the other 3 angles.

Based on the properties of trapezoids, the measure of the other base angle is 30° and the measure of the other diagonal is 10.

The other two angles have a measure of:

$$360 = 30(2) + 2x$$

$$x = 150^\circ$$

The other two angles measure 150° each.

A **right triangle** is a triangle with one right angle. The side opposite the right angle is called the **hypotenuse**. The other two sides are the **legs**. An **altitude** is a line drawn from one vertex, perpendicular to the opposite side.

When an altitude is drawn to the hypotenuse of a right triangle, then the two triangles formed are similar to the original triangle and to each other.

Example:

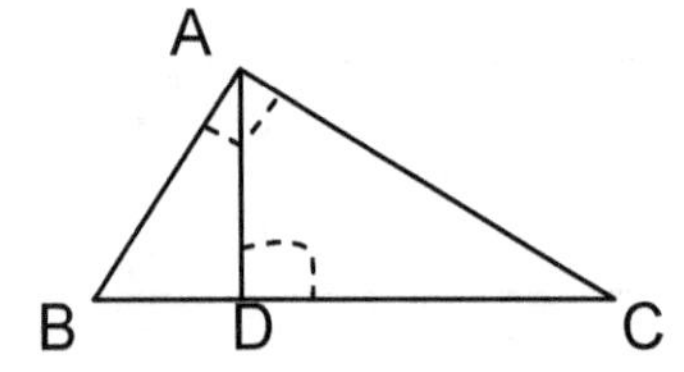

Given right triangle ABC with right angle at A,
altitude AD drawn to hypotenuse BC at D.

$\Delta ABC \sim \Delta ABD \sim \Delta ACD$ The triangles formed are similar to each other and to the original right triangle.

The **Pythagorean theorem** states that the square of the length of the hypotenuse is equal to the sum of the squares of the lengths of the legs. Symbolically, this is stated as:

$$c^2 = a^2 + b^2$$

Given the right triangle below, find the missing side.

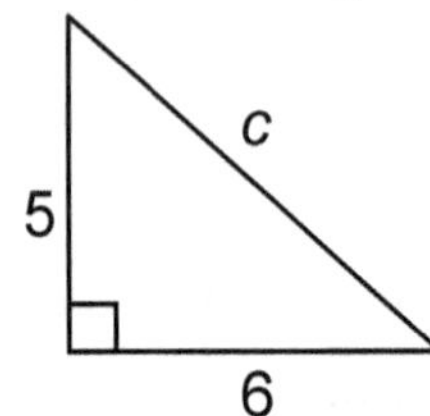

$c^2 = a^2 + b^2$ 1. Write formula
$c^2 = 5^2 + 6^2$ 2.Substitute known values
$c^2 = 61$ 3. Take square root
$c = \sqrt{61}$ or 7.81 4. Solve

Given right triangle ABC, the adjacent side and opposite side can be identified for each angle A and B.

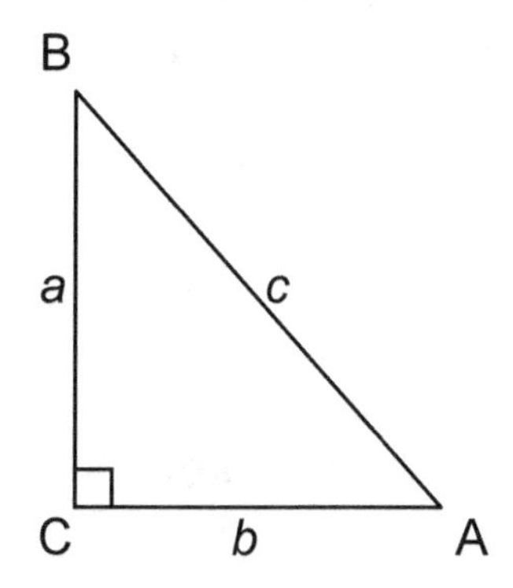

Looking at angle A, it can be determined that side *b* is adjacent to angle A and side *a* is opposite angle A.

If we now look at angle B, we see that side a is adjacent to angle *b* and side *b* is opposite angle B.

The longest side (opposite the 90 degree angle) is always called the hypotenuse.

The basic trigonometric ratios are listed below:

Sine = $\frac{\text{opposite}}{\text{hypotenuse}}$ Cosine = $\frac{\text{adjacent}}{\text{hypotenuse}}$ Tangent = $\frac{\text{opposite}}{\text{adjacent}}$

Sample problem:

1. Use triangle ABC to find the sin, cos and tan for angle A.

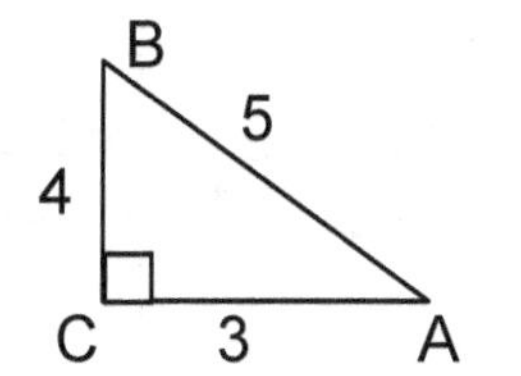

sin A= 4/5
cos A = 3/5
tan A = 4/3

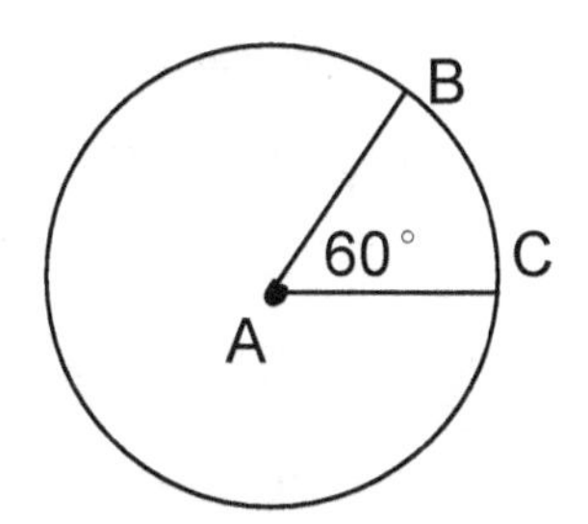

Central angle $BAC = 60^\circ$
Minor arc $BC = 60^\circ$
Major arc $BC = 360 - 60 = 300^\circ$

If you draw **two radii** in a circle, the angle they form with the center as the vertex is a central angle. The piece of the circle "inside" the angle is an arc. Just like a central angle, an arc can have any degree measure from 0 to 360. The measure of an arc is equal to the measure of the central angle which forms the arc. Since a diameter forms a semicircle and the measure of a straight angle like a diameter is 180°, the measure of a semicircle is also 180°.

Given two points on a circle, there are two different arcs which the two points form. Except in the case of semicircles, one of the two arcs will always be greater than 180° and the other will be less than 180°. The arc less than 180° is a minor arc and the arc greater than 180° is a major arc.

Examples:

1.

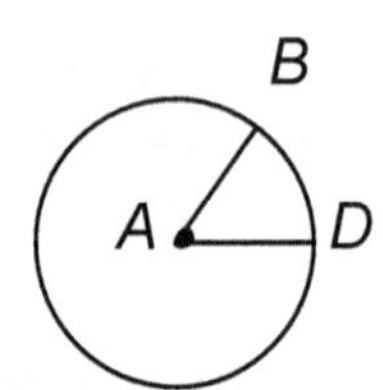

$m\measuredangle BAD = 45^\circ$
What is the measure of the major arc *BD?*

$\measuredangle BAD =$ minor arc *BD*

$45^\circ =$ minor arc *BD*

The measure of the central angle is the same as the measure of the arc it forms.

$360 - 45 =$ major arc *BD*

A major and minor arc always add to 360°.

315° = major arc *BD*

2.

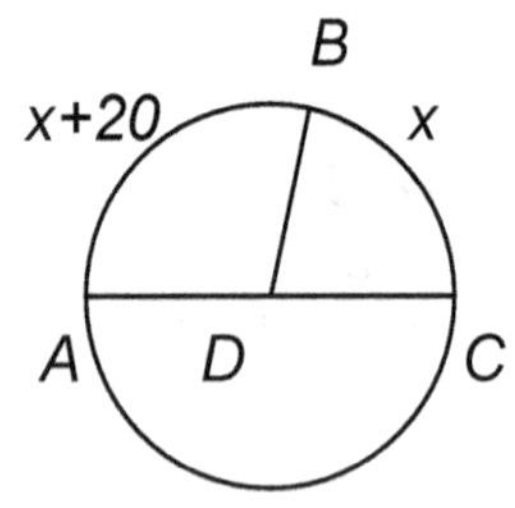

$\overline{AC}$ is a diameter of circle *D*.
What is the measure of $\measuredangle BDC$?

$m\measuredangle ADB + m\measuredangle BDC = 180^\circ$

$x + 20 + x = 180$

$2x + 20 = 180$

$2x = 160$

$x = 80$

A diameter forms a semicircle which has a measure of 180°.

minor arc $BC = 80^\circ$

$m\measuredangle BDC = 80^\circ$

A central angle has the same measure as the arc it forms.

A **tangent line** intersects a circle at exactly one point. If a radius is drawn to that point, the radius will be perpendicular to the tangent.

A chord is a segment with endpoints on the circle. If a radius or diameter is perpendicular to a chord, the radius will cut the chord into two equal parts.

If two chords in the same circle have the same length, the two chords will have arcs that are the same length, and the two chords will be equidistant from the center of the circle. Distance from the center to a chord is measured by finding the length of a segment from the center perpendicular to the chord.

Examples:

1.

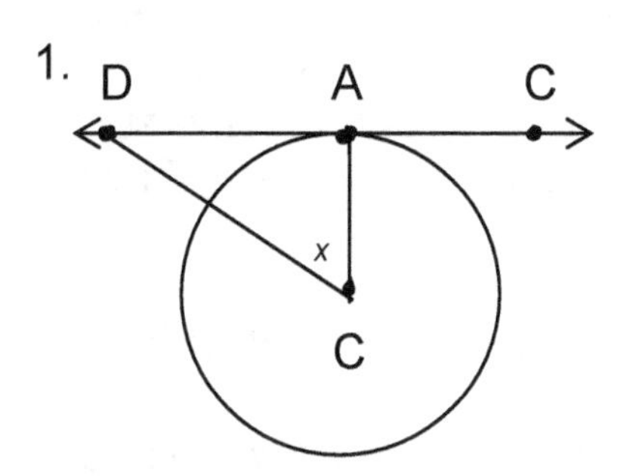

$\overleftrightarrow{DB}$ is tangent to $\odot C$ at A. $m\measuredangle ADC = 40^\circ$. Find x.

$\overline{AC} \perp \overleftrightarrow{DB}$	A radius is $\perp$ to a tangent at the point of tangency.
$m\measuredangle DAC = 90^\circ$	Two segments that are $\perp$ form a 90° angle.
$40 + 90 + x = 180$	The sum of the angles of a triangle is 180°.
$x = 50^\circ$	Solve for x.

2.

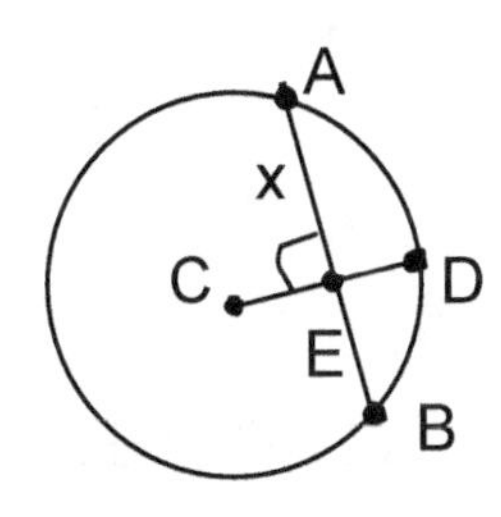

$\overline{CD}$ is a radius and $\overline{CD} \perp$ chord $\overline{AB}$. $\overline{AB} = 10$. Find x.

$$x = \frac{1}{2}(10)$$

$$x = 5$$

If a radius is $\perp$ to a chord, the radius bisects the chord.

Angles with their vertices on the circle:

An inscribed angle is an angle whose vertex is on the circle. Such an angle could be formed by two chords, two diameters, two secants, or a secant and a tangent. An inscribed angle has one arc of the circle in its interior. The measure of the inscribed angle is one-half the measure of this intercepted arc. If two inscribed angles intercept the same arc, the two angles are congruent (i.e. their measures are equal). If an inscribed angle intercepts an entire semicircle, the angle is a right angle.

Angles with their vertices in a circle's interior:

When two chords intersect inside a circle, two sets of vertical angles are formed. Each set of vertical angles intercepts two arcs which are across from each other. The measure of an angle formed by two chords in a circle is equal to one-half the sum of the angle intercepted by the angle and the arc intercepted by its vertical angle.

Angles with their vertices in a circle's exterior:

If an angle has its vertex outside of the circle and each side of the circle intersects the circle, then the angle contains two different arcs. The measure of the angle is equal to one-half the difference of the two arcs.

Examples:

1.

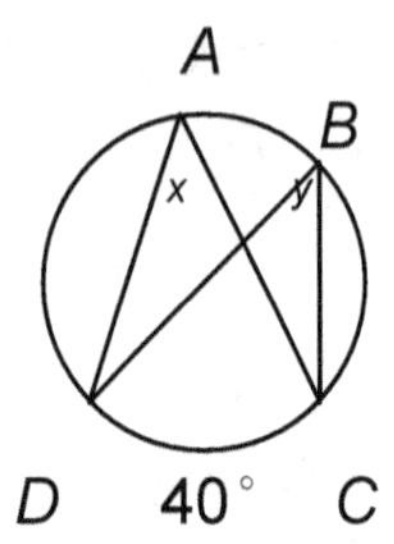

Find *x and y.*
arc $DC = 40^\circ$

$m\measuredangle DAC = \frac{1}{2}(40) = 20^\circ$

$m\measuredangle DBC = \frac{1}{2}(40) = 20^\circ$

$\measuredangle DAC$ and $\measuredangle DBC$ are both inscribed angles, so each one has a measure equal to one-half the measure of arc *DC*.

$x = 20^\circ$ and $y = 20^\circ$

Intersecting chords:

If two chords intersect inside a circle, each chord is divided into two smaller segments. The product of the lengths of the two segments formed from one chord equals the product of the lengths of the two segments formed from the other chord.

Intersecting tangent segments:

If two tangent segments intersect outside a circle, the two segments have the same length.

Intersecting secant segments:

If two secant segments intersect outside a circle, a portion of each segment will lie inside the circle and a portion (called the exterior segment) will lie outside the circle. The product of the length of one secant segment and the length of its exterior segment equals the product of the length of the other secant segment and the length of its exterior segment.

Tangent segments intersecting secant segments:

If a tangent segment and a secant segment intersect outside a circle, the square of the length of the tangent segment equals the product of the length of the secant segment and its exterior segment.

Examples:

1.

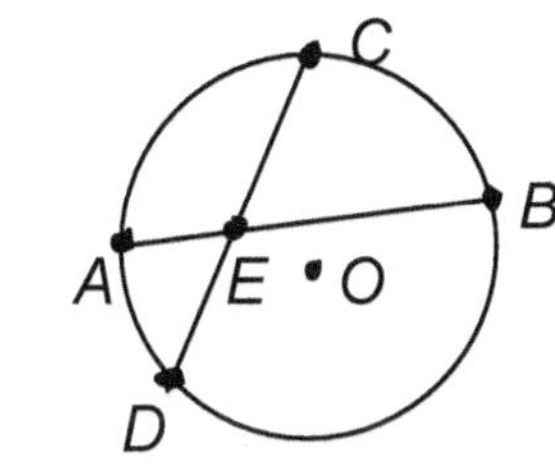

$\overline{AB}$ and $\overline{CD}$ are chords.
$CE=10$, $ED=x$, $AE=5$, $EB=4$

$(AE)(EB)=(CE)(ED)$ — Since the chords intersect in the circle, the products of the segment pieces are equal.

$5(4)=10x$

$20=10x$

$x=2$ — Solve for x.

2.

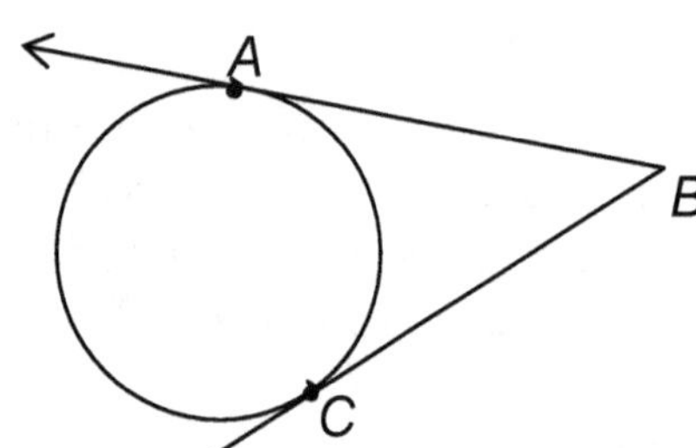

$\overline{AB}$ and $\overline{CD}$ are chords.
$\overline{AB} = x^2 + x - 2$
$\overline{BC} = x^2 - 3x + 5$

Find the length of $\overline{AB}$ and $\overline{BC}$.

$\overline{AB} = x^2 + x - 2$ $\overline{BC} = x^2 - 3x + 5$	Given
$\overline{AB} = \overline{BC}$	Intersecting tangents are equal.
$x^2 + x - 2 = x^2 - 3x + 5$	Set the expression equal and solve.
$4x = 7$ $x = 1.75$	Substitute and solve.
$(1.75)^2 + 1.75 - 2 = \overline{AB}$ $\overline{AB} = \overline{BC} = 2.81$	

Use appropriate problem solving strategies to find the solution.

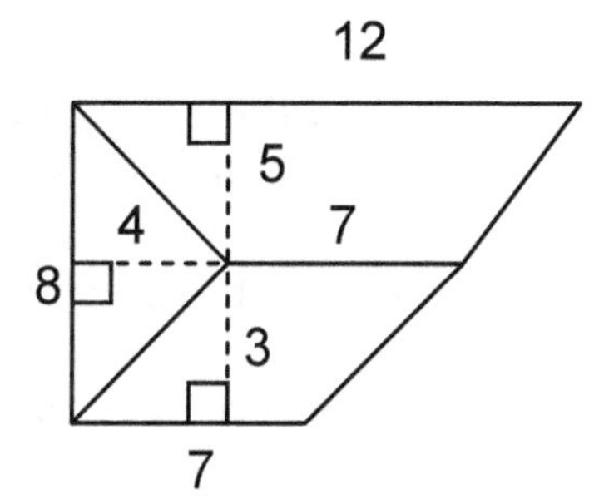

1. Find the area of the given figure.
2. Cut the figure into familiar shapes.
3. Identify what type figures are given and write the appropriate formulas.

Area of figure 1 (triangle)	Area of figure 2 (parallelogram)	Area of figure 3 (trapezoid)
$A=\frac{1}{2}bh$	$A=bh$	$A=\frac{1}{2}h(a+b)$
$A=\frac{1}{2}(8)(4)$	$A=(7)(3)$	$A=\frac{1}{2}(5)(12+7)$
$A=16$ sq. ft	$A=21$ sq. ft	$A=47.5$ sq. ft

Now find the total area by adding the area of all figures.

Total area $=16+21+47.5$
Total area $=84.5$ square ft

Given the figure below, find the area by dividing the polygon into smaller shapes.

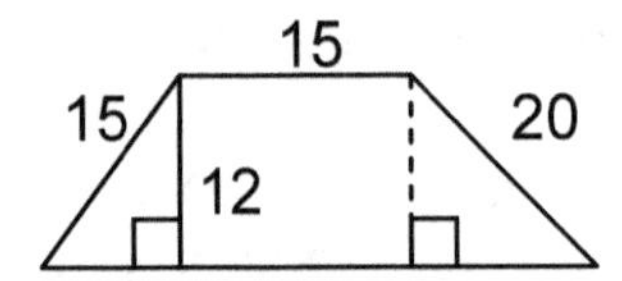

1. Divide the figure into two triangles and a rectangle.
2. Find the missing lengths.
3. Find the area of each part.
4. Find the sum of all areas.

Find base of both right triangles using Pythagorean Formula:

$$a^2 + b^2 = c^2$$
$$a^2 + 12^2 = 15^2$$
$$a^2 = 225 - 144$$
$$a^2 = 81$$
$$a = 9$$

$$a^2 + b^2 = c^2$$
$$a^2 + 12^2 = 20^2$$
$$a^2 = 400 - 144$$
$$a^2 = 256$$
$$a = 16$$

Area of triangle 1	Area of triangle 2	Area of rectangle
$A = \frac{1}{2}bh$	$A = \frac{1}{2}bh$	$A = LW$
$A = \frac{1}{2}(9)(12)$	$A = \frac{1}{2}(16)(12)$	$A = (15)(12)$
$A = 54$ sq. units	$A = 96$ sq. units	$A = 180$ sq. units

Find the sum of all three figures.

$$54 + 96 + 180 = 330 \text{ square units}$$

Competency 0014 Solve problems involving three-dimensional shapes.

FIGURE	LATERAL AREA	TOTAL AREA	VOLUME
Right prism	Ph	2B+Ph	Bh
Regular Pyramid	1/2Pl	1/2Pl+B	1/3Bh

P = Perimeter
h = height
B = Area of Base
l = slant height

Find the total area of the given figure:

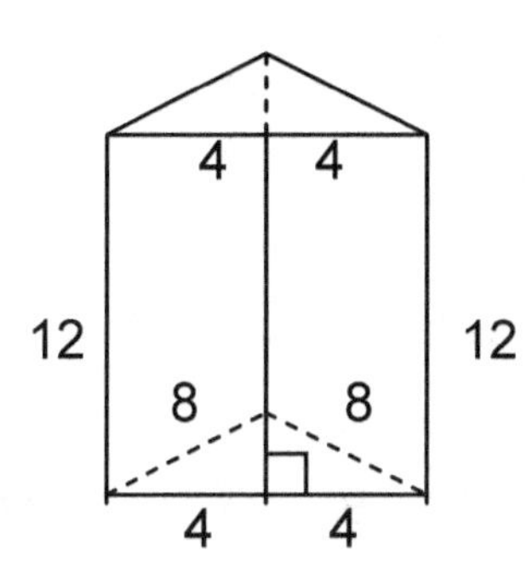

1. Since this is a triangular prism, first find the area of the bases.
2. Find the area of each rectangular lateral face.
3. Add the areas together.

$A=\frac{1}{2}bh$	$A=LW$	1. write formula
$8^2=4^2+h^2$ $h=6.928$		2. find the height of the base triangle
$A=\frac{1}{2}(8)(6.928)$	$A=(8)(12)$	3. substitute known values
$A=27.713$ sq. units	$A=96$ sq. units	4. compute

$$\text{Total Area}=2(27.713)+3(96)$$
$$=343.426 \text{ sq. units}$$

FIGURE	VOLUME	TOTAL SURFACE AREA	LATERAL AREA
Right Cylinder	πr^2h	$2\pi rh+2\pi r^2$	$2\pi rh$
Right Cone	$\frac{\pi r^2h}{3}$	$\pi r\sqrt{r^2+h^2}+\pi r^2$	$\pi r\sqrt{r^2+h^2}$

Note: $\sqrt{r^2+h^2}$ is equal to the slant height of the cone.

Sample problem:

1. A water company is trying to decide whether to use traditional cylindrical paper cups or to offer conical paper cups since both cost the same. The traditional cups are 8 cm wide and 14 cm high. The conical cups are 12 cm wide and 19 cm high. The company will use the cup that holds the most water.

Draw and label a sketch of each.

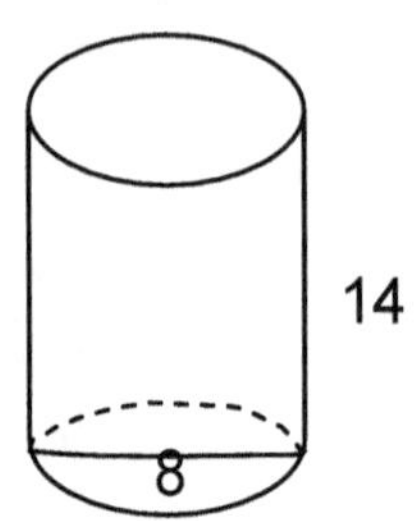

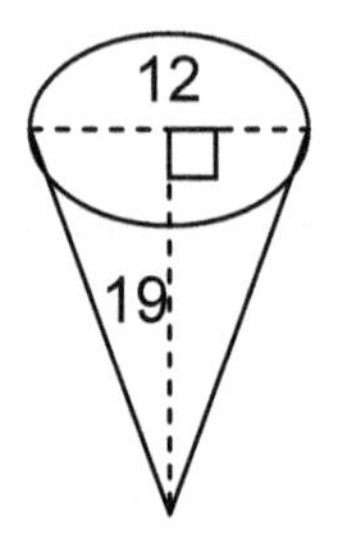

$V = \pi r^2 h$	$V = \dfrac{\pi r^2 h}{3}$	1. Write formula
$V = \pi(4)^2(14)$	$V = \dfrac{1}{3}\pi(6)^2(19)$	2. Substitute
$V = 703.717 \text{ cm}^3$	$V = 716.283 \text{ cm}^3$	3. Solve

The choice should be the conical cup since its volume is greater.

FIGURE	VOLUME	TOTAL SURFACE AREA
Sphere	$\frac{4}{3}\pi r^3$	$4\pi r^2$

Sample problem:

1. How much material is needed to make a basketball that has a diameter of 15 inches? How much air is needed to fill the basketball?

Draw and label a sketch:

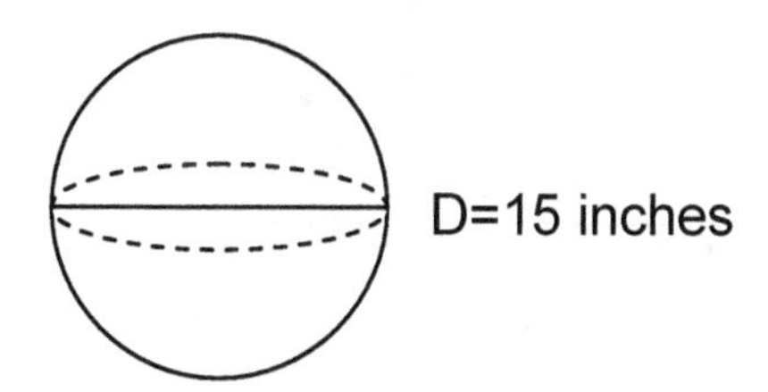

Total surface area	Volume	
$TSA = 4\pi r^2$	$V = \frac{4}{3}\pi r^3$	1. Write formula
$= 4\pi(7.5)^2$	$= \frac{4}{3}\pi(7.5)^3$	2. Substitute
$= 706.8 \text{ in}^2$	$= 1767.1 \text{ in}^3$	3. Solve

PARABOLAS- A parabola is a set of all points in a plane that are equidistant from a fixed point (focus) and a line (directrix).

FORM OF EQUATION	$y = a(x-h)^2 + k$	$x = a(y-k)^2 + h$
IDENTIFICATION	x^2 term, y not squared	y^2 term, x not squared

SKETCH OF GRAPH

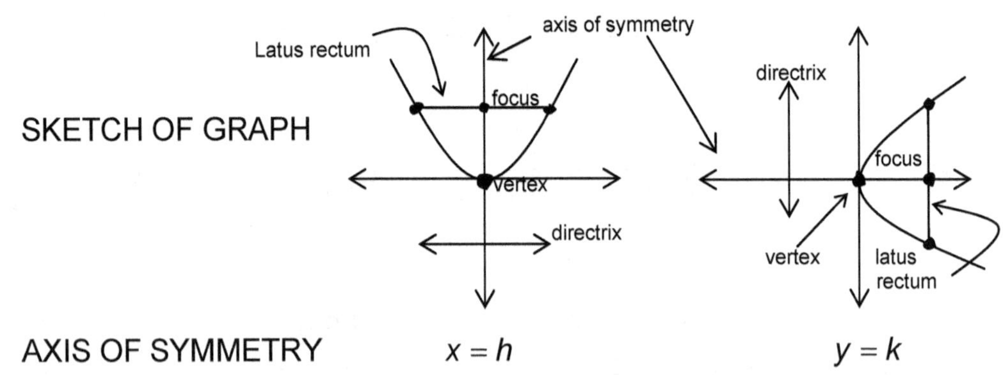

AXIS OF SYMMETRY $x = h$ $y = k$

-A line through the vertex and focus upon which the parabola is symmetric.

VERTEX (h,k) (h,k)

-The point where the parabola intersects the axis of symmetry.

FOCUS	$(h, k + 1/4a)$	$(h + 1/4a, k)$
DIRECTRIX	$y = k - 1/4a$	$x = h - 1/4a$
DIRECTION OF OPENING	up if $a > 0$, down if $a < 0$	right if $a > 0$, left if $a < 0$
LENGTH OF LATUS RECTUM	$\lvert 1/a \rvert$	$\lvert 1/a \rvert$

-A chord through the focus, perpendicular to the axis of symmetry, with endpoints on the parabola.

Sample Problem:

1. Find all identifying features of $y = {}^{-}3x^2 + 6x - 1$.

First, the equation must be put into the general form

$y = a(x-h)^2 + k$.

$y = {}^{-}3x^2 + 6x - 1$ — 1. Begin by completing the square.

$= {}^{-}3(x^2 - 2x + 1) - 1 + 3$

$= {}^{-}3(x-1)^2 + 2$ — 2. Using the general form of the equation to identify known variables.

$a = {}^{-}3 \quad h = 1 \quad k = 2$

axis of symmetry: $x = 1$
vertex: (1,2)
focus: $(1, 1\frac{1}{4})$
directrix: $y = 2\frac{3}{4}$
direction of opening: down since $a < 0$
length of latus rectum: 1/3

ELLIPSE

FORM OF EQUATION	$\frac{(x-h)^2}{a^2}+\frac{(y-k)^2}{b^2}=1$	$\frac{(x-h)^2}{b^2}+\frac{(y-k)^2}{a^2}=1$
(for ellipses where $a^2 > b^2$).	where $b^2 = a^2 - c^2$	where $b^2 = a^2 - c^2$
IDENTIFICATION	horizontal major axis	vertical major axis
SKETCH		

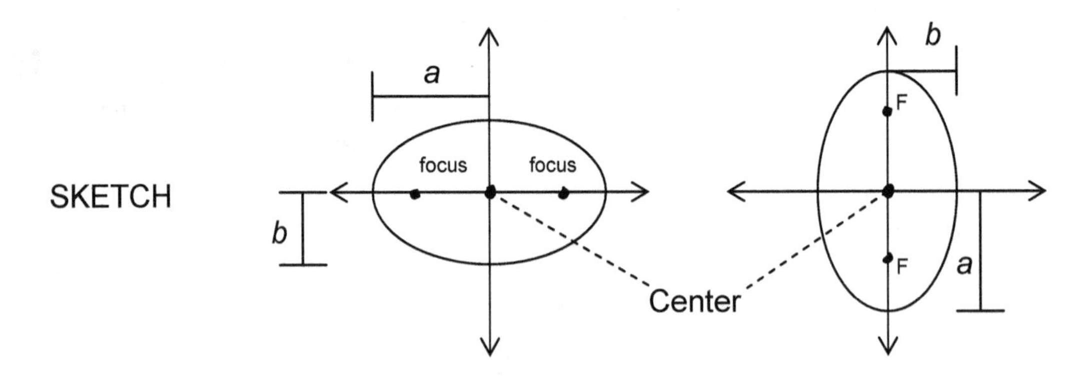

CENTER	(h,k)	(h,k)
FOCI	$(h \pm c, k)$	$(h, k \pm c)$
MAJOR AXIS LENGTH	$2a$	$2a$
MINOR AXIS LENGTH	$2b$	$2b$

Sample Problem:

Find all identifying features of the ellipse $2x^2 + y^2 - 4x + 8y - 6 = 0$.

First, begin by writing the equation in standard form for an ellipse.

$2x^2 + y^2 - 4x + 8y - 6 = 0$ — 1. Complete the square for each variable.

$2(x^2 - 2x + 1) + (y^2 + 8y + 16) = 6 + 2(1) + 16$

$2(x-1)^2 + (y+4)^2 = 24$ — 2. Divide both sides by 24.

$\frac{(x-1)^2}{12} + \frac{(y+4)^2}{24} = 1$ — 3. Now the equation is in standard form.

Identify known variables: $h=1$ $k={}^-4$ $a=\sqrt{24}$ or $2\sqrt{}$
$b=\sqrt{12}$ or $2\sqrt{3}$ $c=2\sqrt{3}$

Identification: vertical major axis

Center: $(1,{}^-4)$

Foci: $(1,{}^-4\pm 2\sqrt{3})$

Major axis: $4\sqrt{6}$

Minor axis: $4\sqrt{3}$

<u>HYPERBOLA</u>

FORM OF EQUATION	$\frac{(x-h)^2}{a^2}-\frac{(y-k)^2}{b^2}=1$ where $c^2=a^2+b^2$	$\frac{(y-k)^2}{a^2}-\frac{(x-h)^2}{b^2}=1$ where $c^2=a^2+b^2$
IDENTIFICATION	horizontal transverse axis (y^2 is negative)	vertical transverse axis (x^2 is negative)
SKETCH	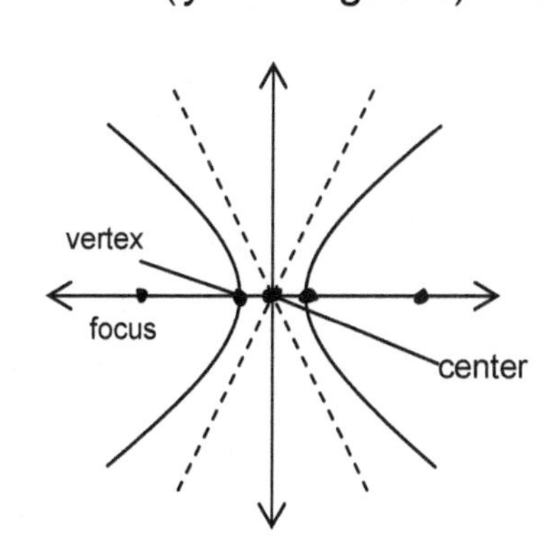	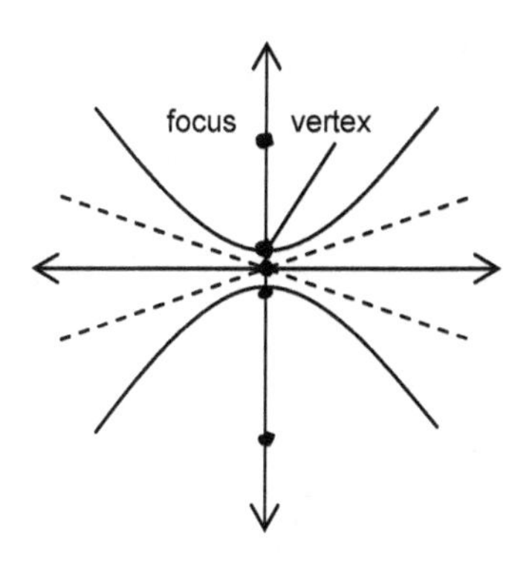
SLOPE OF ASYMPTOTES	$\pm(b/a)$	$\pm(a/b)$
TRANSVERSE AXIS (endpoints are vertices of the hyperbola and go through the center)	$2a$ -on y axis	$2a$ -on x axis
CONJUGATE AXIS (perpendicular to transverse axis at center)	$2b$, -on y axis	$2b$, -on x axis
CENTER	(h,k)	(h,k)
FOCI	$(h\pm c,k)$	$(h,k\pm c)$
VERTICES	$(h\pm a,k)$	$(h,k\pm a)$

Sample Problem:

Find all the identifying features of a hyperbola given its equation.

$$\frac{(x+3)^2}{4}-\frac{(y-4)^2}{16}=1$$

Identify all known variables: $h = {}^-3 \quad k = 4 \quad a = 2 \quad b = 4 \quad c = 2\sqrt{5}$

Slope of asymptotes: $\pm 4/2$ or ± 2
Transverse axis: 4 units long
Conjugate axis: 8 units long
Center: $({}^-3,4)$
Foci: $({}^-3 \pm 2\sqrt{5},4)$
Vertices: $({}^-1,4)$ and $({}^-5,4)$

Sample Problem:

1. Given the equation $x^2 + y^2 = 9$, find the center and the radius of the circle. Then graph the equation.

First, writing the equation in standard circle form gives:

$$(x-0)^2 + (y-0)^2 = 3^2$$

therefore, the center is (0,0) and the radius is 3 units.

Sketch the circle:

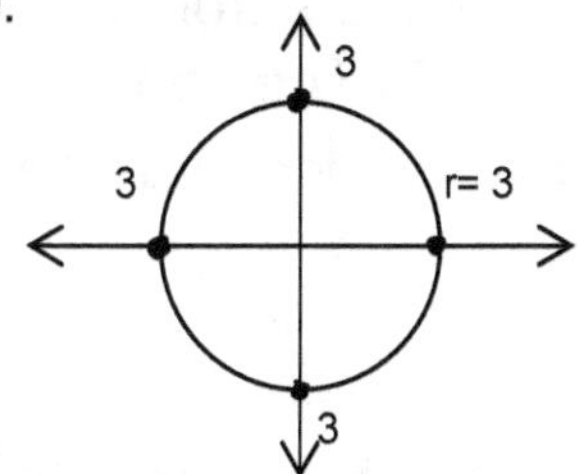

2. Given the equation $x^2 + y^2 - 3x + 8y - 20 = 0$, find the center and the radius. Then graph the circle.

First, write the equation in standard circle form by completing the square for both variables.

$x^2 + y^2 - 3x + 8y - 20 = 0$ 1. Complete the squares.

$(x^2 - 3x + 9/4) + (y^2 + 8y + 16) = 20 + 9/4 + 16$

$(x - 3/2)^2 + (y + 4)^2 = 153/4$

The center is $(3/2, {}^{-}4)$ and the radius is $\dfrac{\sqrt{153}}{2}$ or $\dfrac{3\sqrt{17}}{2}$.

Graph the circle.

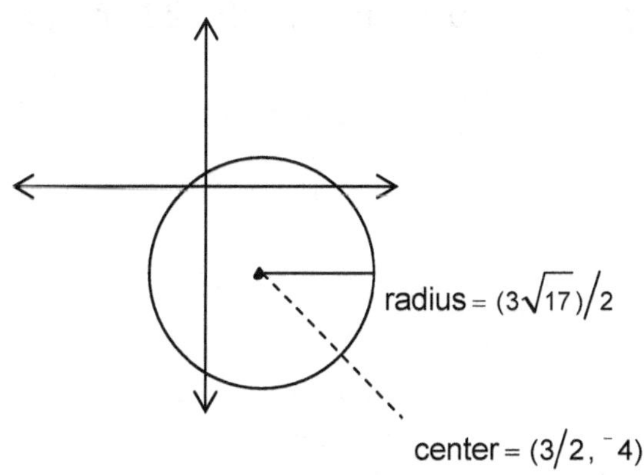

Competency 0015 Understand the principles and properties of coordinate and transformational geometry.

One way to graph points is in the rectangular coordinate system. In this system, the point (a,b) describes the point whose distance along the x-axis is "a" and whose distance along the y-axis is "b." The other method used to locate points is the **polar plane coordinate system**. This system consists of a fixed point called the pole or origin (labeled O) and a ray with O as the initial point called the polar axis. The ordered pair of a point P in the polar coordinate system is (r,θ), where $|r|$ is the distance from the pole and θ is the angle measure from the polar axis to the ray formed by the pole and point P. The coordinates of the pole are $(0,\theta)$, where θ is arbitrary. Angle θ can be measured in either degrees or in radians.

Sample problem:

1. Graph the point P with polar coordinates $(^-2, ^-45 \text{ degrees})$.

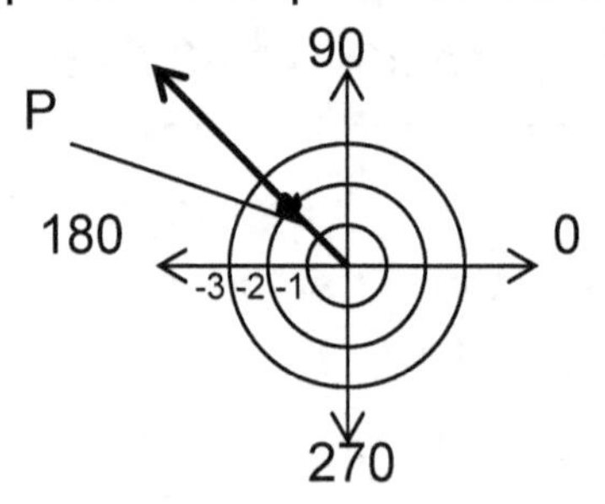

Draw $\theta = {}^-45$ degrees in standard position. Since r is negative, locate the point $|^-2|$ units from the pole on the ray opposite the terminal side of the angle. Note that P can be represented by $(^-2, ^-45 \text{ degrees} + 180 \text{ degrees}) = (2, 135 \text{ degrees})$ or by $(^-2, ^-45 \text{ degrees} - 180 \text{ degrees}) = (2, ^-225 \text{ degrees})$.

2. Graph the point $P = \left(3, \frac{\pi}{4}\right)$ and show another graph that also represents the same point P.

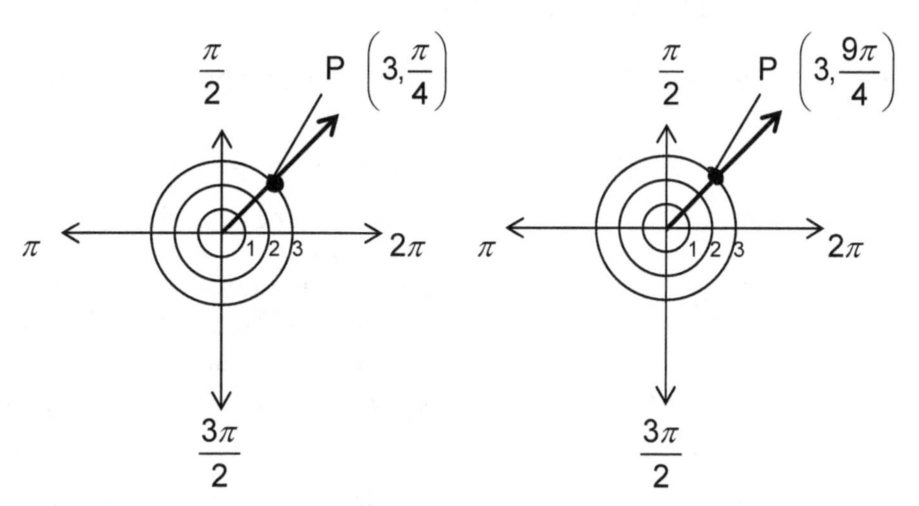

In the second graph, the angle 2π is added to $\frac{\pi}{4}$ to give the point $\left(3, \frac{9\pi}{4}\right)$.

It is possible that r be allowed to be negative. Now instead of measuring $|r|$ units along the terminal side of the angle, we would locate the point $|{}^-3|$ units from the pole on the ray opposite the terminal side. This would give the points $\left({}^-3, \frac{5\pi}{4}\right)$ and $\left({}^-3, \frac{{}^-3\pi}{4}\right)$.

In order to accomplish the task of finding the distance from a given point to another given line the perpendicular line that intersects the point and line must be drawn and the equation of the other line written. From this information the point of intersection can be found. This point and the original point are used in the **distance formula** given below:

$$D = \sqrt{(x_2 - x_1)^2 + (y_2 - y_1)^2}$$

Sample Problem:

1. Given the point $(^{-}4,3)$ and the line $y = 4x + 2$, find the distance from the point to the line.

$y = 4x + 2$	1. Find the slope of the given line by solving for y.
$y = 4x + 2$	2. The slope is 4/1, the perpendicular line will have a slope of $^{-}1/4$.
$y = \left(^{-}1/4\right)x + b$	3. Use the new slope and the given point to find the equation of the perpendicular line.
$3 = \left(^{-}1/4\right)\left(^{-}4\right) + b$	4. Substitute $(^{-}4,3)$ into the equation.
$3 = 1 + b$	5. Solve.
$2 = b$	6. Given the value for b, write the equation of the perpendicular line.
$y = \left(^{-}1/4\right)x + 2$	7. Write in standard form.
$x + 4y = 8$	8. Use both equations to solve by elimination to get the point of intersection.
$\begin{array}{r} ^{-}4x + y = 2 \\ \underline{x + 4y = 8} \end{array}$	9. Multiply the bottom row by 4.
$\begin{array}{r} ^{-}4x + y = 2 \\ \underline{4x + 16y = 32} \\ 17y = 34 \\ y = 2 \end{array}$	10. Solve.
$y = 4x + 2$	11. Substitute to find the x value.
$2 = 4x + 2$ $x = 0$	12. Solve.

(0,2) is the point of intersection. Use this point on the original line and the original point to calculate the distance between them.

$D = \sqrt{(x_2 - x_1)^2 + (y_2 - y_1)^2}$ where points are (0,2) and (-4,3).

$D = \sqrt{(^-4 - 0)^2 + (3 - 2)^2}$ 1. Substitute.

$D = \sqrt{(16) + (1)}$ 2. Simplify.

$D = \sqrt{17}$

The **distance between two parallel lines**, such as line AB and line CD as shown below is the line segment RS, the perpendicular between the two parallels.

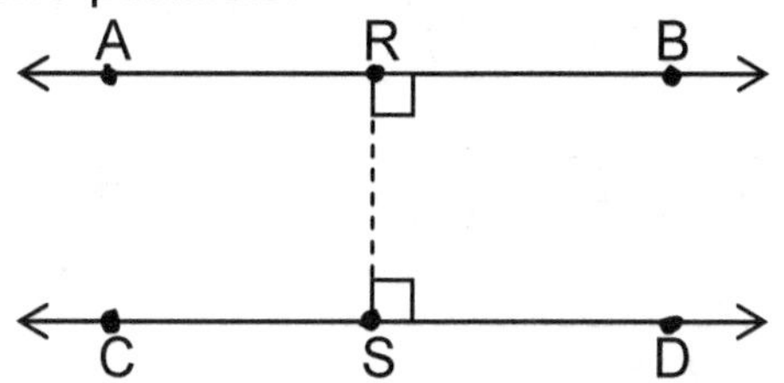

Sample Problem:

Given the geometric figure below, find the distance between the two parallel sides AB and CD.

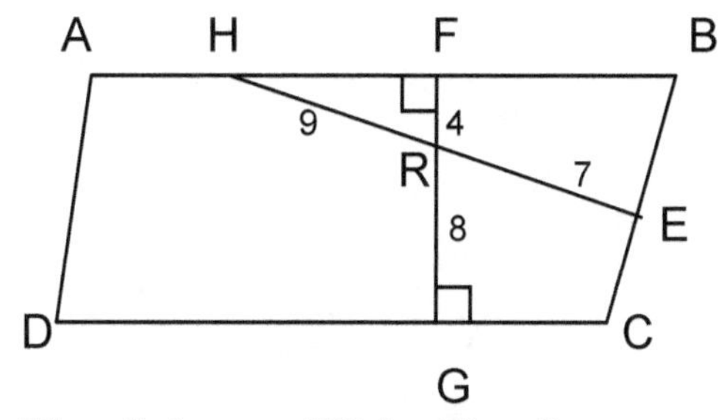

The distance FG is 12 units.

The key to applying the distance formula is to understand the problem before beginning.

$$D = \sqrt{(x_2 - x_1)^2 + (y_2 - y_1)^2}$$

Sample Problem:

1. Find the perimeter of a figure with vertices at (4,5), ($^-4$,6) and ($^-5$, $^-8$).

The figure being described is a triangle. Therefore, the distance for all three sides must be found. Carefully, identify all three sides before beginning.

Side 1 = (4,5) to ($^-4$,6)
Side 2 = ($^-4$,6) to ($^-5$, $^-8$)
Side 3 = ($^-5$, $^-8$) to (4,5)

$$D_1 = \sqrt{(^-4-4)^2+(6-5)^2} = \sqrt{65}$$

$$D_2 = \sqrt{((^-5-(^-4))^2+(^-8-6)^2} = \sqrt{197}$$

$$D_3 = \sqrt{((4-(^-5))^2+(5-(^-8)^2))} = \sqrt{250} \text{ or } 5\sqrt{10}$$

$$\text{Perimeter} = \sqrt{65}+\sqrt{197}+5\sqrt{10}$$

Midpoint Definition:

If a line segment has endpoints of (x_1, y_1) and (x_2, y_2), then the midpoint can be found using:

$$\left(\frac{x_1 + x_2}{2}, \frac{y_1 + y_2}{2}\right)$$

Sample problems:

1. Find the center of a circle with a diameter whose endpoints are (3, 7) and $(-4, -5)$.

$$\text{Midpoint} = \left(\frac{3 + (-4)}{2}, \frac{7 + (-5)}{2}\right)$$

$$\text{Midpoint} = \left(\frac{-1}{2}, 1\right)$$

2. Find the midpoint given the two points $\left(5, 8\sqrt{6}\right)$ and $\left(9, -4\sqrt{6}\right)$.

$$\text{Midpoint} = \left(\frac{5 + 9}{2}, \frac{8\sqrt{6} + (-4\sqrt{6})}{2}\right)$$

$$\text{Midpoint} = \left(7, 2\sqrt{6}\right)$$

SUBAREA IV. DATA ANALYSIS, STATISTICS, AND PROBABILITY

Competency 0016 Understand descriptive statistics and the methods used in collecting, organizing, reporting, and analyzing data.

Percentiles divide data into 100 equal parts. A person whose score falls in the 65th percentile has outperformed 65 percent of all those who took the test. This does not mean that the score was 65 percent out of 100, nor does it mean that 65 percent of the questions answered were correct. It means that the person scored higher than 65 percent of all those who took the test.

Stanine "standard nine" scores combine the understandability of percentages with the properties of the normal curve of probability. Stanines divide the bell curve into nine sections, the largest of which stretches from the 40th to the 60th percentile and is the "Fifth Stanine" (the average of taking into account error possibilities).

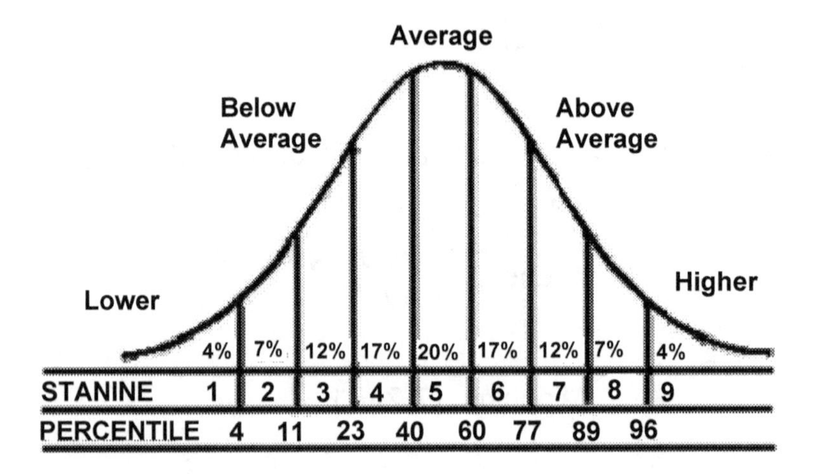

Quartiles divide the data into 4 parts. First find the median of the data set (Q2), then find the median of the upper (Q3) and lower (Q1) halves of the data set. If there are an odd number of values in the data set, include the median value in both halves when finding quartile values. For example, given the data set: {1, 4, 9, 16, 25, 36, 49, 64, 81} first find the median value, which is 25. This is the second quartile. Since there are an odd number of values in the data set (9), we include the median in both halves.

To find the quartile values, we must find the medians of: {1, 4, 9, 16, 25} and {25, 36, 49, 64, 81}. Since each of these subsets had an odd number of elements (5), we use the middle value. Thus the first quartile value is 9 and the third quartile value is 49. If the data set has an even number of elements, average the middle two values. The quartile values are always either one of the data points, or exactly half way between two data points.

Sample problem:

1. Given the following set of data, find the percentile of the score 104.

70, 72, 82, 83, 84, 87, 100, 104, 108, 109, 110, 115

Solution: Find the percentage of scores below 104.

7/12 of the scores are less than 104. This is 58.333%; therefore, the score of 104 is in the 58th percentile.

2. Find the first, second and third quartile for the data listed.

6, 7, 8, 9, 10, 12, 13, 14, 15, 16, 18, 23, 24, 25, 27, 29, 30, 33, 34, 37

Quartile 1: The 1st Quartile is the median of the lower half of the data set, which is 11.

Quartile 2: The median of the data set is the 2nd Quartile, which is 17.

Quartile 3: The 3rd Quartile is the median of the upper half of the data set, which is 28.

Mean, median and mode are three measures of central tendency. The **mean** is the average of the data items. The **median** is found by putting the data items in order from smallest to largest and selecting the item in the middle (or the average of the two items in the middle). The **mode** is the most frequently occurring item.

Range is a measure of variability. It is found by subtracting the smallest value from the largest value.

Sample problem:

Find the mean, median, mode and range of the test scores listed below:

85	77	65
92	90	54
88	85	70
75	80	69
85	88	60
72	74	95

Mean (X) = sum of all scores ÷ number of scores
= 78

Median = put numbers in order from smallest to largest. Pick middle number.

54, 60, 65, 69, 70, 72, 74, 75, 77, 80, 85, 85, 85, 88, 88, 90, 92, 95
-- --
(both in middle)

Therefore, median is average of two numbers in the middle or 78.5

Mode = most frequent number
= 85

Range = largest number minus the smallest number
= 95 − 54
= 41

Different situations require different information. If we examine the circumstances under which an ice cream store owner may use statistics collected in the store, we find different uses for different information.

Over a 7-day period, the store owner collected data on the ice cream flavors sold. He found the mean number of scoops sold was 174 per day. The most frequently sold flavor was vanilla. This information was useful in determining how much ice cream to order in all and in what amounts for each flavor.

In the case of the ice cream store, the median and range had little business value for the owner.

Consider the set of test scores from a math class: 0, 16, 19, 65, 65, 65, 68, 69, 70, 72, 73, 73, 75, 78, 80, 85, 88, and 92. The mean is 64.06 and the median is 71.

Since there are only three scores less than the mean out of the eighteen score, the median (71) would be a more descriptive score.

Retail store owners may be most concerned with the most common dress size so they may order more of that size than any other.

Basic statistical concepts can be applied without computations. For example, inferences can be drawn from a graph of statistical data. A bar graph could display which grade level collected the most money. Student test scores would enable the teacher to determine which units need to be remediated.

Competency 0017 Understand the fundamental principles of probability.

Dependent events occur when the probability of the second event depends on the outcome of the first event. For example, consider the two events (A) it is sunny on Saturday and (B) you go to the beach. If you intend to go to the beach on Saturday, rain or shine, then A and B may be independent. If however, you plan to go to the beach only if it is sunny, then A and B may be dependent. In this situation, the probability of event B will change depending on the outcome of event A.

Suppose you have a pair of dice, one red and one green. If you roll a three on the red die and then roll a four on the green die, we can see that these events do not depend on the other. The total probability of the two independent events can be found by multiplying the separate probabilities.

$$\begin{aligned} P(\text{A and B}) &= P(A) \times P(B) \\ &= 1/6 \times 1/6 \\ &= 1/36 \end{aligned}$$

Many times, however, events are not independent. Suppose a jar contains 12 red marbles and 8 blue marbles. If you randomly pick a red marble, replace it and then randomly pick again, the probability of picking a red marble the second time remains the same. However, if you pick a red marble, and then pick again without replacing the first red marble, the second pick becomes dependent upon the first pick.

$$\begin{aligned} P(\text{Red and Red}) \text{ with replacement} &= P(\text{Red}) \times P(\text{Red}) \\ &= 12/20 \times 12/20 \\ &= 9/25 \end{aligned}$$

$$\begin{aligned} P(\text{Red and Red}) \text{ without replacement} &= P(\text{Red}) \times P(\text{Red}) \\ &= 12/20 \times 11/19 \\ &= 33/95 \end{aligned}$$

Odds are defined as the ratio of the number of favorable outcomes to the number of unfavorable outcomes. The sum of the favorable outcomes and the unfavorable outcomes should always equal the total possible outcomes.

For example, given a bag of 12 red and 7 green marbles, compute the odds of randomly selecting a red marble.

$$\text{Odds of red} = \frac{12}{19}$$

$$\text{Odds of not getting red} = \frac{7}{19}$$

In the case of flipping a coin, it is equally likely that heads or tails will be tossed. The odds of tossing heads are 1:1. This is called even odds.

SUBAREA V. TRIGONOMETRY, CALCULUS, AND DISCRETE MATHEMATICS

Competency 0018 Understand the properties of trigonometric functions and identities.

Unlike trigonometric identities that are true for all values of the defined variable, trigonometric equations are true for some, but not all, of the values of the variable. Most often, trigonometric equations are solved for values between 0 and 360 degrees or 0 and 2π radians.

Some algebraic operations, such as squaring both sides of an equation, will give you extraneous answers. You must remember to check all solutions to be sure that they work.

Sample problems:

1. Solve: $\cos x = 1 - \sin x$ if $0 \leq x < 360$ degrees.

$\cos^2 x = (1 - \sin x)^2$	1. Square both sides
$1 - \sin^2 x = 1 - 2\sin x + \sin^2 x$	2. Substitute
$0 = {}^-2\sin x + 2\sin^2 x$	3. Set = to 0
$0 = 2\sin x({}^-1 + \sin x)$	4. factor
$2\sin x = 0$ ${}^-1 + \sin x = 0$	5. Set each factor = 0
$\sin x = 0$ $\sin x = 1$	6. Solve for $\sin x$
$x = 0$ or 180 $x = 90$	7. Find value of sin at x

The solutions appear to be 0, 90 and 180. Remember to check each solution and you will find that 180 does not give you a true equation. Therefore, the only solutions are 0 and 90 degrees.

2. Solve: $\cos^2 x = \sin^2 x$ if $0 \leq x < 2\pi$

$\cos^2 x = 1 - \cos^2 x$	1. Substitute
$2\cos^2 x = 1$	2. Simplify
$\cos^2 x = \frac{1}{2}$	3. Divide by 2
$\sqrt{\cos^2 x} = \pm\sqrt{\frac{1}{2}}$	4. Take square root
$\cos x = \frac{\pm\sqrt{2}}{2}$	5. Rationalize denominator

$$x = \frac{\pi}{4}, \frac{3\pi}{4}, \frac{5\pi}{4}, \frac{7\pi}{4}$$

Given 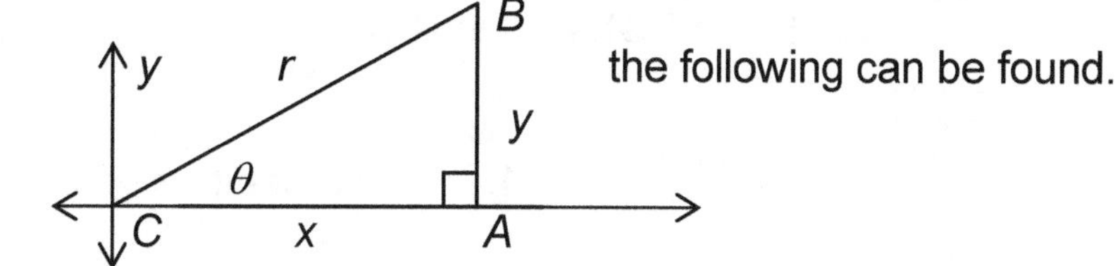 the following can be found.

Trigonometric Functions:

$\sin\theta = \frac{y}{r}$ $\csc\theta = \frac{r}{y}$

$\cos\theta = \frac{x}{r}$ $\sec\theta = \frac{r}{x}$

$\tan\theta = \frac{y}{x}$ $\cot\theta = \frac{x}{y}$

Sample problem:

1. Prove that $\sec\theta = \frac{1}{\cos\theta}$.

$\sec\theta = \frac{1}{\frac{x}{r}}$ Substitute definition of cosine.

$\sec\theta = \frac{1\times r}{\frac{x}{r}\times r}$ Multiply by $\frac{r}{r}$.

$\sec\theta = \frac{r}{x}$ Substitution.

$\sec\theta = \sec\theta$ Substitute definition of $\frac{r}{x}$.

$\sec\theta = \frac{1}{\cos\theta}$ Substitute.

2. Prove that $\sin^2 + \cos^2 = 1$.

$\left(\frac{y}{r}\right)^2 + \left(\frac{x}{r}\right)^2 = 1$	Substitute definitions of sin and cos.
$\frac{y^2 + x^2}{r^2} = 1$	$x^2 + y^2 = r^2$ Pythagorean formula.
$\frac{r^2}{r^2} = 1$	Simplify.
$1 = 1$	Substitute.
$\sin^2\theta + \cos^2\theta = 1$	

Practice problems: Prove each identity.

1. $\cot\theta = \frac{\cos\theta}{\sin\theta}$ 2. $1 + \cot^2\theta = \csc^2\theta$

There are two methods that may be used to prove trigonometric identities. One method is to choose one side of the equation and manipulate it until it equals the other side. The other method is to replace expressions on both sides of the equation with equivalent expressions until both sides are equal.

The Reciprocal Identities

$\sin x = \frac{1}{\csc x}$	$\sin x \csc x = 1$	$\csc x = \frac{1}{\sin x}$
$\cos x = \frac{1}{\sec x}$	$\cos x \sec x = 1$	$\sec x = \frac{1}{\cos x}$
$\tan x = \frac{1}{\cot x}$	$\tan x \cot x = 1$	$\cot x = \frac{1}{\tan x}$
$\tan x = \frac{\sin x}{\cos x}$		$\cot x = \frac{\cos x}{\sin x}$

The Pythagorean Identities

$\sin^2 x + \cos^2 x = 1$ $1 + \tan^2 x = \sec^2 x$ $1 + \cot^2 x = \csc^2 x$

Sample problems:

1. Prove that $\cot x + \tan x = (\csc x)(\sec x)$.

$$\frac{\cos x}{\sin x} + \frac{\sin x}{\cos x}$$ Reciprocal identities.

$$\frac{\cos^2 x + \sin^2 x}{\sin x \cos x}$$ Common denominator.

$$\frac{1}{\sin x \cos x}$$ Pythagorean identity.

$$\frac{1}{\sin x} \times \frac{1}{\cos x}$$

$$\csc x(\sec x) = \csc x(\sec x)$$ Reciprocal identity, therefore,

$$\cot x + \tan x = \csc x(\sec x)$$

2. Prove that $\dfrac{\cos^2 \theta}{1 + 2\sin\theta + \sin^2 \theta} = \dfrac{\sec\theta - \tan\theta}{\sec\theta + \tan\theta}$.

$$\frac{1 - \sin^2 \theta}{(1 + \sin\theta)(1 + \sin\theta)} = \frac{\sec\theta - \tan\theta}{\sec\theta + \tan\theta}$$ Pythagorean identity

factor denominator.

$$\frac{1 - \sin^2 \theta}{(1 + \sin\theta)(1 + \sin\theta)} = \frac{\frac{1}{\cos\theta} - \frac{\sin\theta}{\cos\theta}}{\frac{1}{\cos\theta} + \frac{\sin\theta}{\cos\theta}}$$ Reciprocal identities.

$$\frac{(1 - \sin\theta)(1 + \sin\theta)}{(1 + \sin\theta)(1 + \sin\theta)} = \frac{\frac{1 - \sin\theta}{\cos\theta}(\cos\theta)}{\frac{1 + \sin\theta}{\cos\theta}(\cos\theta)}$$ Factor $1 - \sin^2 \theta$.

Multiply by $\dfrac{\cos\theta}{\cos\theta}$.

$$\frac{1 - \sin\theta}{1 + \sin\theta} = \frac{1 - \sin\theta}{1 + \sin\theta}$$ Simplify.

$$\frac{\cos^2 \theta}{1 + 2\sin\theta + \sin^2 \theta} = \frac{\sec\theta - \tan\theta}{\sec\theta + \tan\theta}$$

It is easiest to graph trigonometric functions when using a calculator by making a table of values.

DEGREES

	0	30	45	60	90	120	135	150	180	210	225	240	270	300	315	330	360
sin	0	.5	.71	.87	1	.87	.71	.5	0	-.5	-.71	-.87	-1	-.87	-.71	-.5	0
cos	1	.87	.71	.5	0	-.5	-.71	-.87	-1	-.87	-.71	-.5	0	.5	.71	.87	1
tan	0	.58	1	1.7	--	-1.7	-1	-.58	0	.58	1	1.7	--	-1.7	-1	-.58	0
	0	$\frac{\pi}{6}$	$\frac{\pi}{4}$	$\frac{\pi}{3}$	$\frac{\pi}{2}$	$\frac{2\pi}{3}$	$\frac{3\pi}{4}$	$\frac{5\pi}{6}$	π	$\frac{7\pi}{6}$	$\frac{5\pi}{4}$	$\frac{4\pi}{3}$	$\frac{3\pi}{2}$	$\frac{5\pi}{3}$	$\frac{7\pi}{4}$	$\frac{11\pi}{6}$	2π

RADIANS

Remember the graph always ranges from $+1$ to $^{-}1$ for sine and cosine functions unless noted as the coefficient of the function in the equation. For example, $y = 3\cos x$ has an amplitude of 3 units from the center line (0). Its maximum and minimum points would be at $+3$ and $^{-}3$.

Tangent is not defined at the values 90 and 270 degrees or $\frac{\pi}{2}$ and $\frac{3\pi}{2}$. Therefore, vertical asymptotes are drawn at those values.

The inverse functions can be graphed in the same manner using a calculator to create a table of values.

In order to **solve a right triangle** using trigonometric functions it is helpful to identify the given parts and label them. Usually more than one trigonometric function may be appropriately applied.

Some items to know about right triangles:

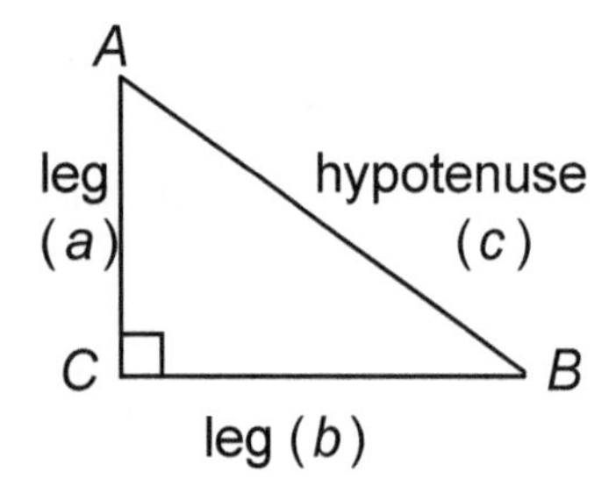

Given angle A, the side labeled leg (a) Is adjacent to angle A. And the side(b) is opposite to angle A.

Sample problem:

1. Find the missing side.

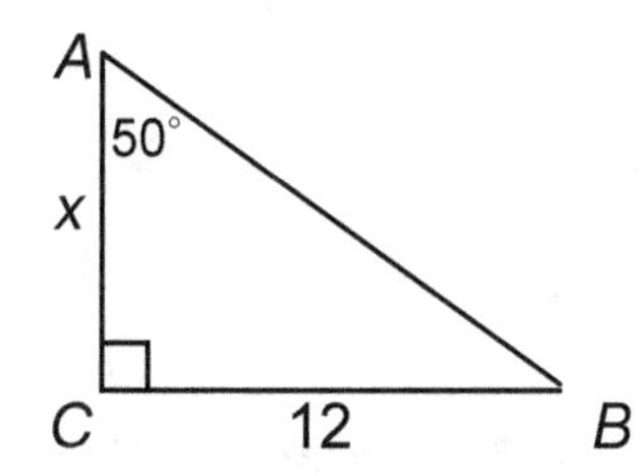

1. Identify the known values. Angle $A = 50$ degrees. The side opposite the given angle is 12. The missing side is the adjacent leg.
2. The information suggests the use of the tangent function
3. Write the function.
4. Substitute.
5. Solve.

$$\tan A = \frac{\text{opposite}}{\text{adjacent}}$$

$$\tan 50 = \frac{12}{x}$$

$$1.192 = \frac{12}{x}$$

$$x(1.192) = 12$$

$$x = 10.069$$

Remember that since angle A and angle B are complimentary, then angle $B = 90 - 50$ or 40 degrees.

Using this information we could have solved for the same side only this time it is the leg opposite from angle B.

1. Write the formula.
2. Substitute.
3. Solve.

$$\tan B = \frac{\text{opposite}}{\text{adjacent}}$$

$$\tan 40 = \frac{x}{12}$$

$$12(.839) = x$$

$$10.069 \approx x$$

Now that the two sides of the triangle are known, the third side can be found using the Pythagorean Theorem.

Definition: For any triangle ABC, when given two sides and the included angle, the other side can be found using one of the formulas below:

$$a^2 = b^2 + c^2 - (2bc)\cos A$$
$$b^2 = a^2 + c^2 - (2ac)\cos B$$
$$c^2 = a^2 + b^2 - (2ab)\cos C$$

Similarly, when given three sides of a triangle, the included angles can be found using the derivation:

$$\cos A = \frac{b^2 + c^2 - a^2}{2bc}$$
$$\cos B = \frac{a^2 + c^2 - b^2}{2ac}$$
$$\cos C = \frac{a^2 + b^2 - c^2}{2ab}$$

Sample problem:

1. Solve triangle ABC, if angle $B = 87.5^\circ$, $a = 12.3$, and $c = 23.2$. (Compute to the nearest tenth).

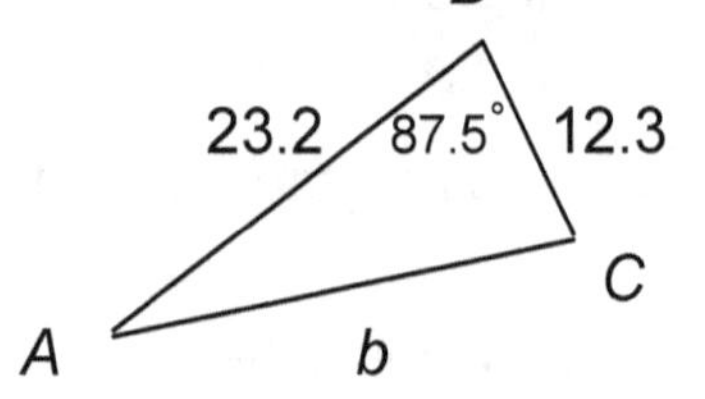

1. Draw and label a sketch.

Find side b.

$b^2 = a^2 + c^2 - (2ac)\cos B$ — 2. Write the formula.

$b^2 = (12.3)^2 + (23.2)^2 - 2(12.3)(23.2)(\cos 87.5)$ — 3. Substitute.

$b^2 = 664.636$

$b = 25.8$ (rounded) — 4. Solve.

Use the law of sines to find angle A.

$\frac{\sin A}{a} = \frac{\sin B}{b}$ — 1. Write formula.

$\frac{\sin A}{12.3} = \frac{\sin 87.5}{25.8} = \frac{12.29}{25.8}$ — 2. Substitute.

$\sin A = 0.47629$ — 3. Solve.
Angle $A = 28.4$

Therefore, angle $C = 180 - (87.5 + 28.4)$

2. Solve triangle ABC if $a = 15$, $b = 21$, and $c = 18$. (Round to the nearest tenth).

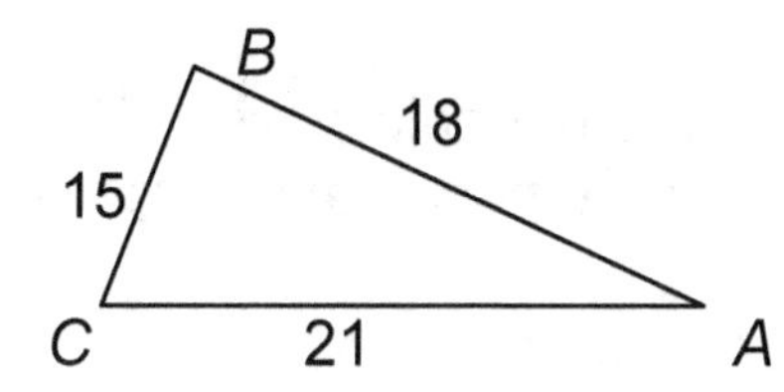

1. Draw and label a sketch.

Find angle A.

$\cos A = \frac{b^2 + c^2 - a^2}{2bc}$ — 2. Write formula.

$\cos A = \frac{21^2 + 18^2 - 15^2}{2(21)(18)}$ — 3. Substitute.

$\cos A = 0.714$ — 4. Solve.
Angle $A = 44.4$

Find angle B.

$\cos B = \frac{a^2 + c^2 - b^2}{2ac}$ — 5. Write formula.

$\cos B = \frac{15^2 + 18^2 - 21^2}{2(15)(18)}$ — 6. Substitute.

$\cos B = 0.2$ — 7. Solve.
Angle $B = 78.5$

Therefore, angle $C = 180 - (44.4 + 78.5)$
$= 57.1$

The trigonometric functions sine, cosine, and tangent are periodic functions. The values of periodic functions repeat on regular intervals. Period, amplitude, and phase shift are key properties of periodic functions that can be determined by observation of the graph.

The **period** of a function is the smallest domain containing the complete cycle of the function. For example, the period of a sine or cosine function is the distance between the peaks of the graph.

The **amplitude** of a function is half the distance between the maximum and minimum values of the function.

Phase shift is the amount of horizontal displacement of a function from its original position.

Both sine and cosine graphs are periodic waves. Below is a generic sine/cosine graph with the period and amplitude labeled.

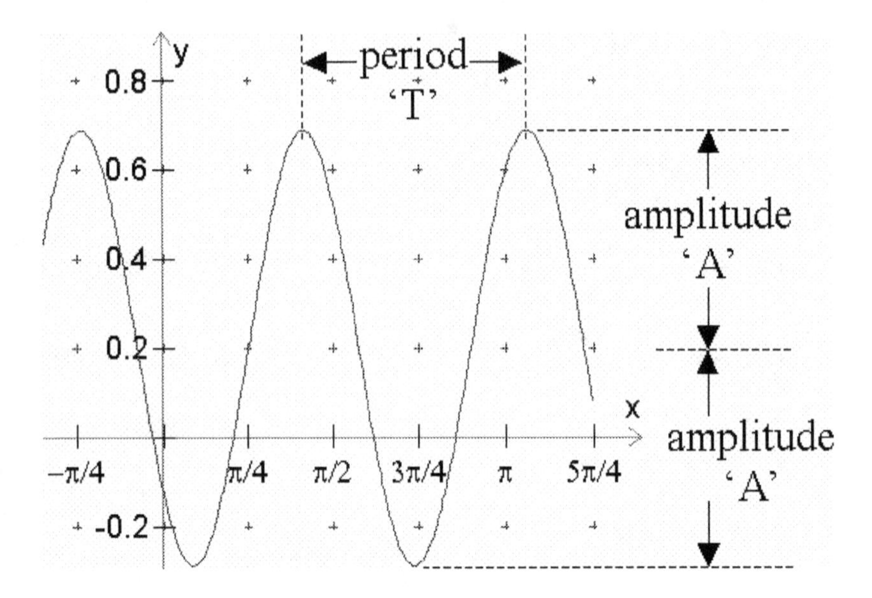

Properties of the graphs of basic trigonometric functions:

Function	Period	Amplitude
y = sin x	2π radians	1
y = cos x	2π radians	1
y = tan x	π radians	undefined

Below are the graphs of the basic trigonometric functions, (a) y = sin x; (b) y = cos x; and (c) y= tan x.

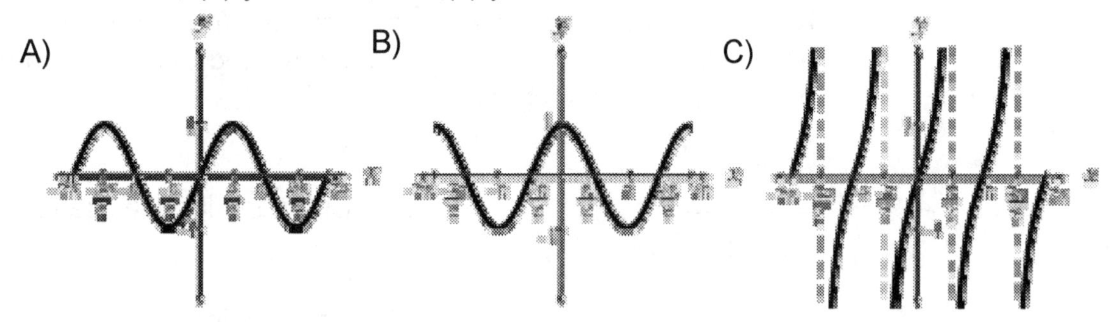

Note that the phase shift of trigonometric graphs is the horizontal distance displacement of the curve from these basic functions.
Definition: For any triangle ABC, where a, b, and c are the lengths of the sides opposite angles A, B, and C respectively.

$$\frac{\sin A}{a}=\frac{\sin B}{b}=\frac{\sin C}{c}$$

Sample problem:

1. An inlet is 140 feet wide. The lines of sight from each bank to an approaching ship are 79 degrees and 58 degrees. What are the distances from each bank to the ship?

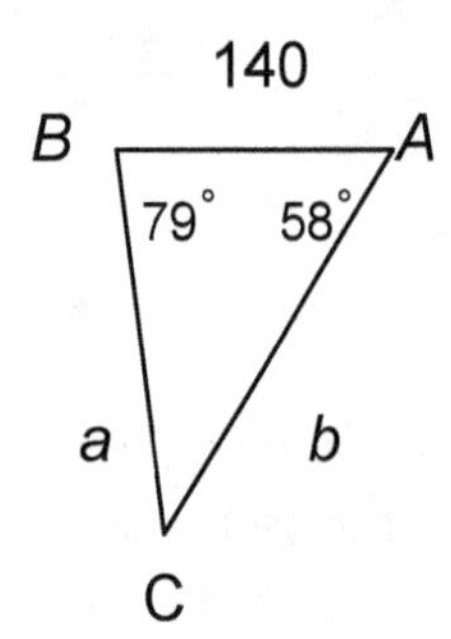

1. Draw and label a sketch.

2. The missing angle is
$180-(79+58)=43$
$\measuredangle C=43$ degrees.

$$\frac{\sin A}{a}=\frac{\sin B}{b}=\frac{\sin C}{c}$$

3. Write formula.

Side opposite 79 degree angle:

$$\frac{\sin 79}{b}=\frac{\sin 43}{140}$$

4. Substitute.

$$b=\frac{140(.9816)}{.6820}$$

5. Solve.

$$b\approx 201.501 \text{ feet}$$

Side opposite 58 degree angle:

$$\frac{\sin 58}{a}=\frac{\sin 43}{140}$$

6. Substitute.

$$a=\frac{140(.848)}{.6820}$$

7. Solve.

$$a\approx 174.076 \text{ feet}$$

Competency 0019 Understand the conceptual basis of calculus.

The limit of a function is the y value that the graph approaches as the x values approach a certain number. To find a limit there are two points to remember.

1. Factor the expression completely and cancel all common factors in fractions.
2. Substitute the number to which the variable is approaching. In most cases this produces the value of the limit.

If the variable in the limit is approaching ∞, factor and simplify first; then examine the result. If the result does not involve a fraction with the variable in the denominator, the limit is usually also equal to ∞. If the variable is in the denominator of the fraction, the denominator is getting larger which makes the entire fraction smaller. In other words the limit is zero.

Examples:

1. $\lim_{x \to {}^-3} \frac{x^2+5x+6}{x+3} + 4x$ Factor the numerator.

$\lim_{x \to {}^-3} \frac{(x+3)(x+2)}{(x+3)} + 4x$ Cancel the common factors.

$\lim_{x \to {}^-3} (x+2) + 4x$ Substitute $^-3$ for x.

$({}^-3+2)+4({}^-3)$ Simplify.

${}^-1 + {}^-12$

${}^-13$

2. $\lim_{x \to \infty} \frac{2x^2}{x^5}$ Cancel the common factors.

$\lim_{x \to \infty} \frac{2}{x^3}$

Since the denominator is getting larger, the entire fraction is getting smaller. The fraction is getting close to zero.

$\frac{2}{\infty^3}$

Practice problems:

1. $\lim_{x \to \pi} 5x^2 + \sin x$

2. $\lim_{x \to -4} = \frac{x^2+9x+20}{x+4}$

After simplifying an expression to evaluate a limit, substitute the value that the variable approaches. If the substitution results in either $0/0$ or ∞/∞, use L'Hopital's rule to find the limit.

L'Hopital's rule states that you can find such limits by taking the derivative of the numerator and the derivative of the denominator, and then finding the limit of the resulting quotient.

Examples:

1. $\lim_{x\to\infty} \frac{3x-1}{x^2+2x+3}$ No factoring is possible.

$\frac{3\infty-1}{\infty^2+2\infty+3}$ Substitute ∞ for x.

$\frac{\infty}{\infty}$ Since a constant times infinity is still a large number, $3(\infty)=\infty$.

$\lim_{x\to\infty} \frac{3}{2x+2}$ To find the limit, take the derivative of the numerator and denominator.

$\frac{3}{2(\infty)+2}$ Substitute ∞ for x again.

$\frac{3}{\infty}$

0 Since the denominator is a very large number, the fraction is getting smaller. Thus the limit is zero.

2. $\lim_{x \to 1} \frac{\ln x}{x-1}$ — Substitute 1 for x.

$\frac{\ln 1}{1-1}$ — The $\ln 1 = 0$

$\frac{0}{0}$ — To find the limit, take the derivative of the numerator and denominator.

$\lim_{x \to 1} \frac{\frac{1}{x}}{1}$ — Substitute 1 for x again.

$\frac{\frac{1}{1}}{1}$ — Simplify. The limit is one.

1

Practice problems:

1. $\lim_{x \to \infty} \frac{x^2 - 3}{x}$

2. $\lim_{x \to \frac{\pi}{2}} \frac{\cos x}{x - \frac{\pi}{2}}$

The derivative of a function has two basic interpretations.

I. Instantaneous rate of change
II. Slope of a tangent line at a given point

If a question asks for the rate of change of a function, take the derivative to find the equation for the rate of change. Then plug in for the variable to find the instantaneous rate of change.

The following is a list summarizing some of the more common quantities referred to in rate of change problems.

area	height	profit
decay	population growth	sales
distance	position	temperature
frequency	pressure	volume

Pick a point, say $x = {}^{-}3$, on the graph of a function. Draw a tangent line at that point. Find the derivative of the function and plug in $x = {}^{-}3$. The result will be the slope of the tangent line.

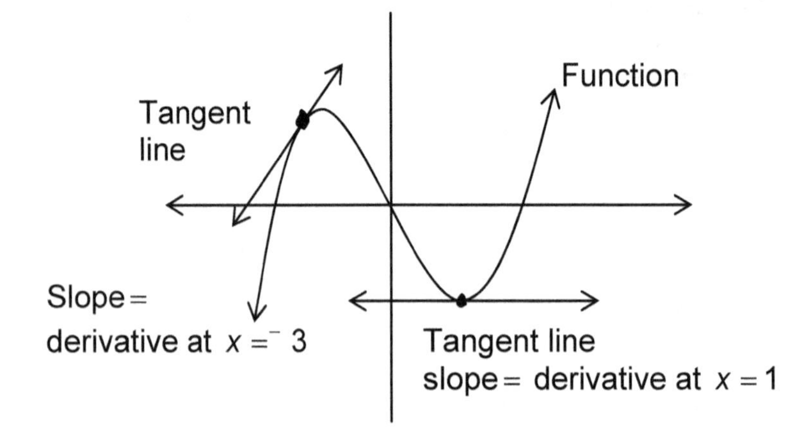

A function is said to be increasing if it is rising from left to right and decreasing if it is falling from left to right. Lines with positive slopes are increasing, and lines with negative slopes are decreasing. If the function in question is something other than a line, simply refer to the slopes of the tangent lines as the test for increasing or decreasing. Take the derivative of the function and plug in an x value to get the slope of the tangent line; a positive slope means the function is increasing and a negative slope means it is decreasing. If an interval for x values is given, just pick any point between the two values to substitute.

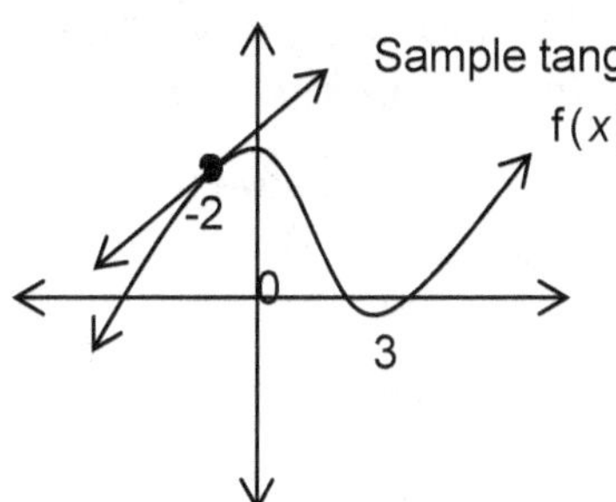

On the interval $(^-2,0)$, $f(x)$ is increasing. The tangent lines on this part of the graph have positive slopes.

Example:

The growth of a certain bacteria is given by $f(x) = x + \frac{1}{x}$. Determine if the rate of growth is increasing or decreasing on the time interval $(^-1,0)$.

$f\,'(x) = 1 + \frac{^-1}{x^2}$

To test for increasing or decreasing, find the slope of the tangent line by taking the derivative.

$f\,'\left(\frac{^-1}{2}\right) = 1 + \frac{^-1}{(^-1/2)^2}$

Pick any point on $(^-1,0)$ and substitute into the derivative.

$f\,'\left(\frac{^-1}{2}\right) = 1 + \frac{^-1}{1/4}$

$= 1 - 4$

$= {^-3}$

The slope of the tangent line at $x = \frac{^-1}{2}$ is $^-3$. The exact value of the slope is not important. The important fact is that the slope is negative.

Substituting an x value into a function produces a corresponding y value. The coordinates of the point (x, y), where y is the largest of all the y values, is said to be a **maximum point**. The coordinates of the point (x, y), where y is the smallest of all the y values, is said to be a **minimum point**. To find these points, only a few x values must be tested. First, find all of the x values that make the derivative either zero or undefined. Substitute these values into the original function to obtain the corresponding y values. Compare the y values. The largest y value is a maximum; the smallest y value is a minimum. If the question asks for the maxima or minima on an interval, be certain to also find the y values that correspond to the numbers at either end of the interval.

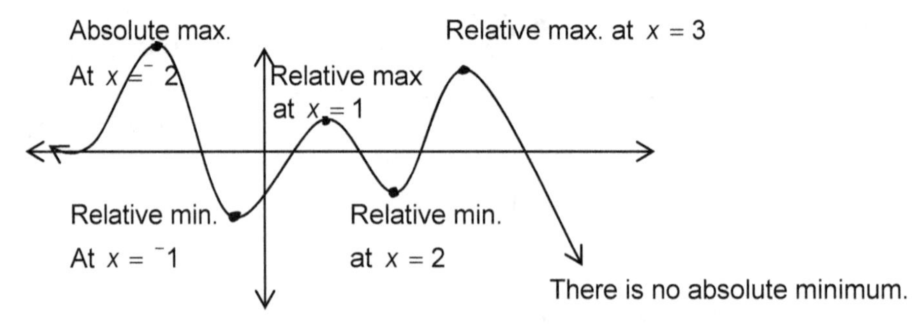

Example:

Find the maxima and minima of $f(x) = 2x^4 - 4x^2$ at the interval $(^-2, 1)$.

$f'(x) = 8x^3 - 8x$	Take the derivative first. Find all the x values (critical values) that
$8x^3 - 8x = 0$	make the derivative zero or undefined. In this case, there are no x values that make the
$8x(x^2 - 1) = 0$	derivative undefined.
$8x(x-1)(x+1) = 0$	Substitute the critical values into
$x = 0,\ x = 1, \text{ or } x = {}^-1$	the original function. Also, plug
$f(0) = 2(0)^4 - 4(0)^2 = 0$	in the endpoint of the interval.
$f(1) = 2(1)^4 - 4(1)^2 = {}^-2$	Note that 1 is a critical point
$f(^-1) = 2(^-1)^4 - 4(^-1)^2 = {}^-2$	and an endpoint.
$f(^-2) = 2(^-2)^4 - 4(^-2)^2 = 16$	

The maximum is at (–2, 16) and there are minima at (1, –2) and (–1, –2). (0,0) is neither the maximum or minimum on (–2, 1) but it is still considered a relative extra point.

The first derivative reveals whether a curve is rising or falling **(increasing or decreasing)** from the left to the right. In much the same way, the second derivative relates whether the curve is concave up or concave down. Curves which are concave up are said to "collect water;" curves which are concave down are said to "dump water." To find the intervals where a curve is concave up or concave down, follow the following steps.

1. Take the second derivative (i.e. the derivative of the first derivative).
2. Find the critical x values.
   -Set the second derivative equal to zero and solve for critical x values.
   -Find the x values that make the second derivative undefined (i.e. make the denominator of the second derivative equal to zero).
   Such values may not always exist.
3. Pick sample values which are both less than and greater than each of the critical values.
4. Substitute each of these sample values into the second derivative and determine whether the result is positive or negative.
   -If the sample value yields a positive number for the second derivative, the curve is concave up on the interval where the sample value originated.
   -If the sample value yields a negative number for the second derivative, the curve is concave down on the interval where the sample value originated.

Example:

Find the intervals where the curve is concave up and concave down for $f(x) = x^4 - 4x^3 + 16x - 16$.

$f'(x) = 4x^3 - 12x^2 + 16$ Take the second derivative.

$f''(x) = 12x^2 - 24x$

$12x^2 - 24x = 0$ Find the critical values by setting the second derivative equal to zero.

$12x(x-2) = 0$ There are no values that make the second derivative undefined.

$x = 0$ or $x = 2$

Set up a number line with the critical values.

0 2

Sample values: $^-1$, 1, 3

Pick sample values in each of the 3 intervals. If the sample value produces a negative number, the function is concave down.

$f''(^-1) = 12(^-1)^2 - 24(^-1) = 36$
$f''(1) = 12(1)^2 - 24(1) = ^-12$
$f''(3) = 12(3)^2 - 24(3) = 36$

If the value produces a positive number, the curve is concave up. If the value produces a zero, the function is linear.

Therefore when $x < 0$ the function is concave up,
when $0 < x < 2$ the function is concave down,
when $x > 2$ the function is concave up.

A **point of inflection** is a point where a curve changes from being concave up to concave down or vice versa. To find these points, follow the steps for finding the intervals where a curve is concave up or concave down. A critical value is part of an inflection point if the curve is concave up on one side of the value and concave down on the other. The critical value is the x coordinate of the inflection point. To get the y coordinate, plug the critical value into the original function.

Example: Find the inflection points of $f(x) = 2x - \tan x$ where $\frac{^-\pi}{2} < x < \frac{\pi}{2}$.

$(x) = 2x - \tan x \qquad \frac{^-\pi}{2} < x < \frac{\pi}{2}$ — Note the restriction on x.

$f'(x) = 2 - \sec^2 x$ — Take the second derivative.

$f''(x) = 0 - 2 \bullet \sec x \bullet (\sec x \tan x)$ — Use the Power rule.

$= ^-2 \bullet \frac{1}{\cos x} \bullet \frac{1}{\cos x} \bullet \frac{\sin x}{\cos x}$ — The derivative of $\sec x$ is $(\sec x \tan x)$.

$f''(x) = \frac{^-2\sin x}{\cos^3 x}$ — Find critical values by solving for the second derivative equal to zero.

$0 = \frac{^-2\sin x}{\cos^3 x}$ — No x values on $\left(\frac{^-\pi}{2}, \frac{\pi}{2}\right)$ make the denominator zero.

$$^{-}2\sin x = 0$$
$$\sin x = 0$$
$$x = 0$$

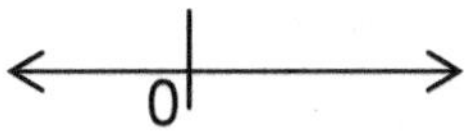

Pick sample values on each side of the critical value $x = 0$.

Sample values: $x = \frac{^{-}\pi}{4}$ and $x = \frac{\pi}{4}$

$$f''\left(\frac{^{-}\pi}{4}\right) = \frac{^{-}2\sin(^{-}\pi/4)}{\cos^3(\pi/4)} = \frac{^{-}2(^{-}\sqrt{2}/2)}{(\sqrt{2}/2)^3} = \frac{\sqrt{2}}{(\sqrt{8}/8)} = \frac{8\sqrt{2}}{\sqrt{8}} = \frac{8\sqrt{2}}{\sqrt{8}} \cdot \frac{\sqrt{8}}{\sqrt{8}}$$
$$= \frac{8\sqrt{16}}{8} = 4$$

$$f''\left(\frac{\pi}{4}\right) = \frac{^{-}2\sin(\pi/4)}{\cos^3(\pi/4)} = \frac{^{-}2(\sqrt{2}/2)}{(\sqrt{2}/2)^3} = \frac{^{-}\sqrt{2}}{(\sqrt{8}/8)} = \frac{^{-}8\sqrt{2}}{\sqrt{8}} = -4$$

The second derivative is positive on $(0,\infty)$ and negative on $(^{-}\infty,0)$. So the curve changes concavity at $x = 0$. Use the original equation to find the y value that inflection occurs at.

$f(0) = 2(0) - \tan 0 = 0 - 0 = 0$

The inflection point is (0,0).

Extreme value problems are also known as max-min problems. Extreme value problems require using the first derivative to find values which either maximize or minimize some quantity such as area, profit, or volume. Follow these steps to solve an extreme value problem.

1. Write an equation for the quantity to be maximized or minimized.
2. Use the other information in the problem to write secondary equations.
3. Use the secondary equations for substitutions, and rewrite the original equation in terms of only one variable.
4. Find the derivative of the primary equation (step 1) and the critical values of this derivative.
5. Substitute these critical values into the primary equation.

The value which produces either the largest or smallest value is used to find the solution.

Example:

A manufacturer wishes to construct an open box from the piece of metal shown below by cutting squares from each corner and folding up the sides. The square piece of metal is 12 feet on a side. What are the dimensions of the squares to be cut out which will maximize the volume?

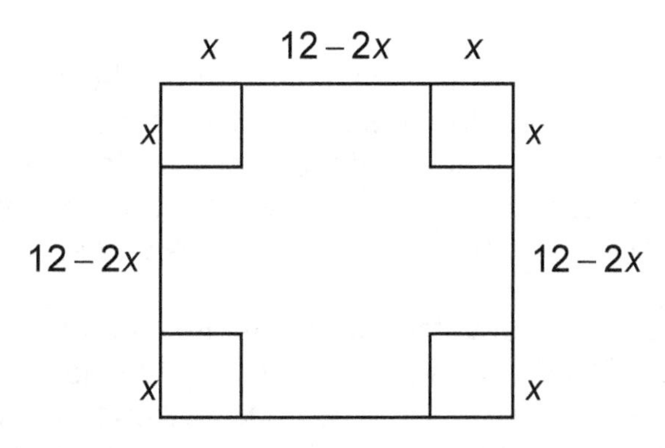

$\text{Volume} = lwh$	Primary equation.
$l = 12 - 2x$	
$w = 12 - 2x$	Secondary equations.
$h = x$	
$V = (12 - 2x)(12 - 2x)(x)$	Make substitutions.
$V = (144x - 48x^2 + 4x^3)$	Take the derivative.
$\frac{dV}{dx} = 144 - 96x + 12x^2$	
$0 = 12(x^2 - 8x + 12)$	Find critical values by setting the derivative equal to zero.
$0 = 12(x - 6)(x - 2)$	
$x = 6$ and $x = 2$	Substitute critical values into volume equation.
$V = 144(6) - 48(6)^2 + 4(6)^3$	$V = 144(2) - 48(2)^2 + 4(2)^3$
$V = 0 \text{ ft}^3$ when $x = 6$	$V = 128 \text{ ft}^3$ when $x = 2$

Therefore, the manufacturer can maximize the volume if the squares to be cut out are 2 feet by 2 feet ($x = 2$).

If a particle (or a car, a bullet, etc.) is moving along a line, then the distance that the particle travels can be expressed by a function in terms of time.

1. The first derivative of the distance function will provide the velocity function for the particle. Substituting a value for time into this expression will provide the instantaneous velocity of the particle at the time.

   Velocity is the rate of change of the distance traveled by the particle. Taking the absolute value of the derivative provides the speed of the particle. A positive value for the velocity indicates that the particle is moving forward, and a negative value indicates the particle is moving backwards.

2. The second derivative of the distance function (which would also be the first derivative of the velocity function) provides the acceleration function. The acceleration of the particle is the rate of change of the velocity. If a value for time produces a positive acceleration, the particle is speeding up; if it produces a negative value, the particle is slowing down. If the acceleration is zero, the particle is moving at a constant speed.

To find the time when a particle stops, set the first derivative (i.e. the velocity function) equal to zero and solve for time. This time value is also the instant when the particle changes direction.

Example:

The motion of a particle moving along a line is according to the equation: $s(t) = 20 + 3t - 5t^2$ where s is in meters and t is in seconds.

Find the position, velocity, and acceleration of a particle at $t = 2$ seconds.

$s(2) = 20 + 3(2) - 5(2)^2$ $= 6$ meters	Plug $t = 2$ into the original equation to find the position.
$s'(t) = v(t) = 3 - 10t$	The derivative of the first function gives the velocity.
$v(2) = 3 - 10(2) = {}^{-}17$ m/s	Plug $t = 2$ into the velocity function to find the velocity. ${}^{-}17$ m/s indicates the particle is moving backwards.
$s''(t) = a(t) = {}^{-}10$	The second derivation of position gives the acceleration.
$a(2) = {}^{-}10$ m/s²	Substituting $t = 2$, yields an acceleration of ${}^{-}10$ m/s², which indicates the particle is slowing down.

Finding the **rate of change** of one quantity (for example distance, volume, etc.) with respect to time is often referred to as a rate of change problem. To find an instantaneous rate of change of a particular quantity, write a function in terms of time for that quantity; then take the derivative of the function. Substitute the values at which the instantaneous rate of change is sought.

Functions which are in terms of more than one variable may be used to find related rates of change. These functions are often not written in terms of time. To find a related rate of change, follow these steps.

1. Write an equation which relates all the quantities referred to in the problem.
2. Take the derivative of both sides of the equation with respect to time.
   Follow the same steps as used in implicit differentiation. This means take the derivative of each part of the equation remembering to multiply each term by the derivative of the variable involved with respect to time. For example, if a term includes the variable v for volume, take the derivative of the term remembering to multiply by dv/dt for the derivative of volume with respect to time. dv/dt is the rate of change of the volume.
3. Substitute the known rates of change and quantities, and solve for the desired rate of change.

Example:

1. What is the instantaneous rate of change of the area of a circle where the radius is 3 cm?

$A(r) = \pi r^2$	Write an equation for area.
$A'(r) = 2\pi r$	Take the derivative to find the rate of change.
$A'(3) = 2\pi(3) = 6\pi$	Substitute in $r = 3$ to arrive at the instantaneous rate of change.

An integral is almost the same thing as an antiderivative: the only difference is the notation.

$\int_{-2}^{1} 2x\,dx$ is the integral form of the antiderivative of $2x$. The numbers at the top and bottom of the integral sign (1 and $^-2$) are the numbers used to find the exact value of this integral. If these numbers are used the integral is said to be *definite* and does not have an unknown constant c in the answer.

The fundamental theorem of calculus states that an integral such as the one above is equal to the antiderivative of the function inside (here $2x$) evaluated from $x = {}^-2$ to $x = 1$. To do this, follow these steps.

1. Take the antiderivative of the function inside the integral.
2. Plug in the upper number (here $x = 1$) and plug in the lower number (here $x = {}^-2$), giving two expressions.
3. Subtract the second expression from the first to achieve the integral value.

Examples:

1. $\int_{-2}^{1} 2x\,dx = x^2\Big]_{-2}^{1}$ — Take the antiderivative.

   $\int_{-2}^{1} 2x\,dx = 1^2 - ({}^-2)^2$ — Substitute in $x = 1$ and $x = {}^-2$ and subtract the results.

   $\int_{-2}^{1} 2x\,dx = 1 - 4 = {}^-3$ — The integral has the value $^-3$.

2. $\int_{0}^{\pi/2} \cos x\,dx = \sin x\Big]_{0}^{\pi/2}$ — The antiderivative of $\cos x$ is $\sin x$.

   $\int_{0}^{\frac{\pi}{2}} \cos x\,dx = \sin\frac{\pi}{2} - \sin 0$ — Substitute in $x = \frac{\pi}{2}$ and $x = 0$. Subtract the results.

   $\int_{0}^{\frac{\pi}{2}} \cos x\,dx = 1 - 0 = 1$ — The integral has the value 1.

Taking the integral of a function and evaluating it from one x value to another provides the **total area under the curve** (i.e. between the curve and the x axis). Remember, though, that regions above the x axis have "positive" area and regions below the x axis have "negative" area. You must account for these positive and negative values when finding the area under curves. Follow these steps.

1. Determine the x values that will serve as the left and right boundaries of the region.
2. Find all x values between the boundaries that are either solutions to the function or are values which are not in the domain of the function. These numbers are the interval numbers.
3. Integrate the function.
4. Evaluate the integral once for each of the intervals using the boundary numbers.
5. If any of the intervals evaluates to a negative number, make it positive (the negative simply tells you that the region is below the x axis).
6. Add the value of each integral to arrive at the area under the curve.

Example:

Find the area under the following function on the given intervals.
$f(x) = \sin x$; $(0, 2\pi)$

$\sin x = 0$ — Find any roots to f(x) on $(0, 2\pi)$.

$x = \pi$

$(0, \pi)$ $(\pi, 2\pi)$ — Determine the intervals using the boundary numbers and the roots.

$\int \sin x dx = {}^{-}\cos x$ — Integrate f(x). We can ignore the constant c because we have numbers to use to evaluate.

$${}^{-}\cos x\Big]_{x=0}^{x=\pi} = {}^{-}\cos \pi - ({}^{-}\cos 0)$$

$${}^{-}\cos x\Big]_{x=0}^{x=\pi} = {}^{-}(-1) + (1) = 2$$

$${}^{-}\cos x\Big]_{x=\pi}^{x=2\pi} = {}^{-}\cos 2\pi - ({}^{-}\cos \pi)$$

$${}^{-}\cos x\Big]_{x=\pi}^{x=2\pi} = {}^{-}1 + ({}^{-}1) = {}^{-}2$$

The ${}^{-}2$ means that for $(\pi, 2\pi)$, the region is below the x axis, but the area is still 2.

Area $= 2 + 2 = 4$ — Add the 2 integrals together to get the area.

Competency 0020 Understand the principles of discrete/finite mathematics.

Sequences can be **finite** or **infinite**. A finite sequence is a sequence whose domain consists of the set {1, 2, 3, ... *n*} or the first *n* positive integers. An infinite sequence is a sequence whose domain consists of the set {1, 2, 3, ...}; which is, in other words, all positive integers.

A **recurrence relation** is an equation that defines a sequence recursively; in other words, each term of the sequence is defined as a function of the preceding terms.

A real-life application would be using a recurrence relation to determine how much your savings would be in an account at the end of a certain period of time. For example:

You deposit $5,000 in your savings account. Your bank pays 5% interest compounded annually. How much will your account be worth at the end of 10 years?

Let V represent the amount of money in the account and V_n represent the amount of money after n years.

The amount in the account after n years equals the amount in the account after n – 1 years plus the interest for the nth year. This can be expressed as the recurrence relation V_0 where your initial deposit is represented by $V_0 = 5,000$.

$$V_0 = V_0$$
$$V_1 = 1.05V_0$$
$$V_2 = 1.05V_1 = (1.05)^2 V_0$$
$$V_3 = 1.05V_2 = (1.05)^3 V_0$$
$$\ldots\ldots$$
$$V_n = (1.05)V_{n-1} = (1.05)^n V_0$$

Inserting the values into the equation, you get $V_{10} = (1.05)^{10}(5,000) = 8,144$.

You determine that after investing $5,000 in an account earning 5% interest, compounded annually for 10 years, you would have $8,144.

Graphs display data so that the data can be interpreted. Graphs are often used to see trends and predict future performance.
For example, this line graph depicts the auto sales for a car dealership. The car dealership is able to see at a glance how many cars were sold in a particular month and which months tended to have the least and greatest sales. This information helps him to control his inventory, forecast his sales, and manage his staffing. He might also use the information to plan ways in which to boost sales in lagging months.

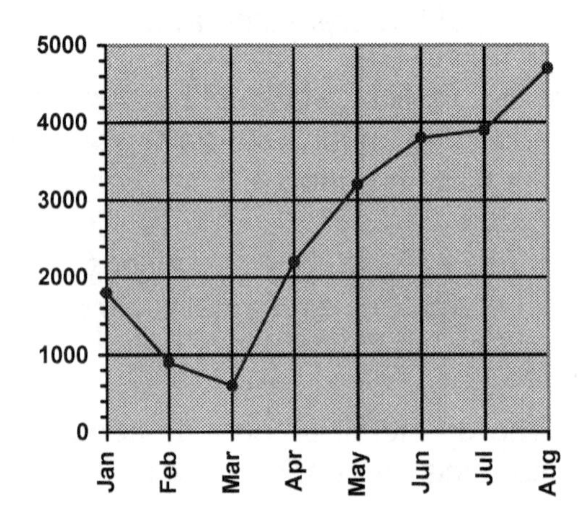

A **matrix** is an ordered set of numbers in rectangular form.

$$\begin{pmatrix} 0 & 3 & 1 \\ 4 & 2 & 3 \\ 1 & 0 & 2 \end{pmatrix}$$

Since this matrix has 3 rows and 3 columns, it is called a 3 x 3 matrix. The element in the second row, third column would be denoted as $3_{2,3}$.

Matrices are used often to solve systems of equations. They are also used by physicists, mathematicians, and biologists to organize and study data such as population growth. They are also used in finance for such purposes as investment growth analysis and portfolio analysis. Matrices are easily translated into computer code in high-level programming languages and can be easily expressed in electronic spreadsheets.

A simple financial example of using a matrix to solve a problem follows:

A company has two stores. The income and expenses (in dollars) for the two stores, for three months, are shown in the matrices.

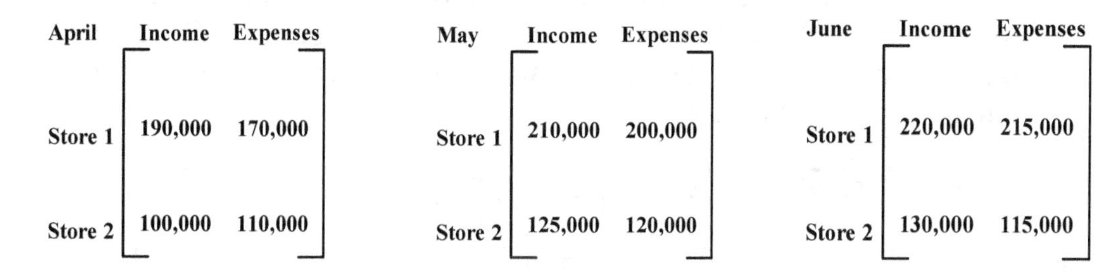

April	Income	Expenses
Store 1	190,000	170,000
Store 2	100,000	110,000

May	Income	Expenses
Store 1	210,000	200,000
Store 2	125,000	120,000

June	Income	Expenses
Store 1	220,000	215,000
Store 2	130,000	115,000

The owner wants to know what his first-quarter income and expenses were, so he adds the three matrices.

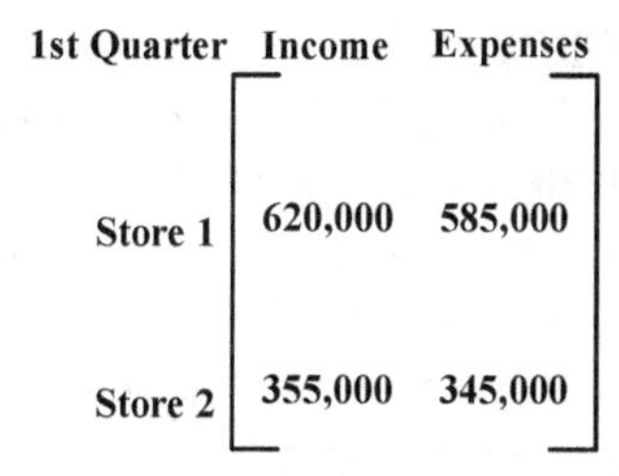

1st Quarter	Income	Expenses
Store 1	620,000	585,000
Store 2	355,000	345,000

Then, to find the profit for each store:

Profit for Store 1 = $620,000 - $585,000 = $35,000
Profit for Store 2 = $355,000 - $345,000 = $10,000

An **algorithm** is a method of calculating; simply put, it can be multiplication, subtraction, or a combination of operations. When working with computers and calculators we employ **algorithmic thinking**, which means performing mathematical tasks by creating a sequential and often repetitive set of steps. A simple example would be to create an algorithm to generate the Fibonacci numbers utilizing the MR and M+ keys found on most calculators. The table below shows Entry made in the calculator, the value *x* seen in the display, and the value M contained in the memory.

Entry	ON/AC	1	M+	+	M+	MR	+	M+	MR	+	...
x	0	1	1	1	1	2	3	3	5	8	...
M	0	0	1	1	2	2	2	5	5	5	...

This eliminates the need to repeatedly enter required numbers. Computers have to be programmed, and many advanced calculators are programmable. A **program** is a series of steps of an algorithm that are entered into a computer or calculator. The main advantage of using a program is that once the algorithm is entered, a result may be obtained by merely hitting a single keystroke to select the program, thereby eliminating the need to continually enter a large number of steps. Teachers find that programmable calculators are excellent for investigating "what if?" situations. Using graphing calculators or computer software has many advantages. The technology is better able to handle large data sets, such as the results of a science experiment, and it is much easier to edit and sort the data and change the style of the graph to find its best representation. Furthermore, graphing calculators also provide a tool to plot statistics.

Answer Key to Practice Problems

Competency 0007, *page 23*

Question #2 a, b, c, f are functions

Question #3 Domain = $^-\infty, \infty$ Range = $^-5, \infty$

Question #4 Domain = $^-\infty, \infty$ Range = $^-6, \infty$

Question #5 Domain = 1,4,7,6 Range = -2

Question #6 Domain = $x \neq 2, ^-2$

Question #7 Domain = $^-\infty, \infty$ Range = -4, 4

Domain = $^-\infty, \infty$ Range = $2, \infty$

Question #8 Domain = $^-\infty, \infty$ Range = 5

Question #9 (3,9), (-4,16), (6,3), (1,9), (1,3)

Competency 0008, *page 26*

Question #1

x > 3

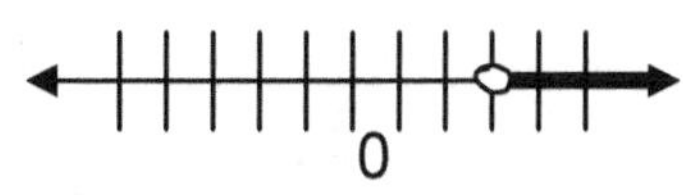

Question #2

x = 2

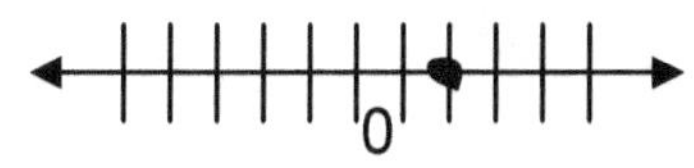

Question #3

$x \leq 6$

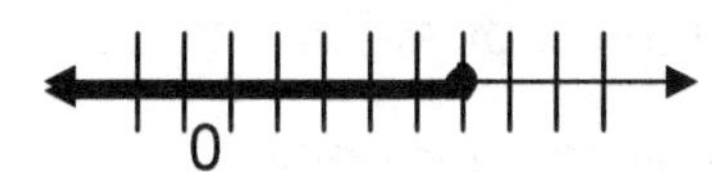

Question #4

x = -4

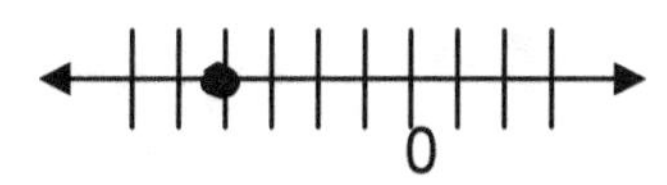

page 28

Question #1

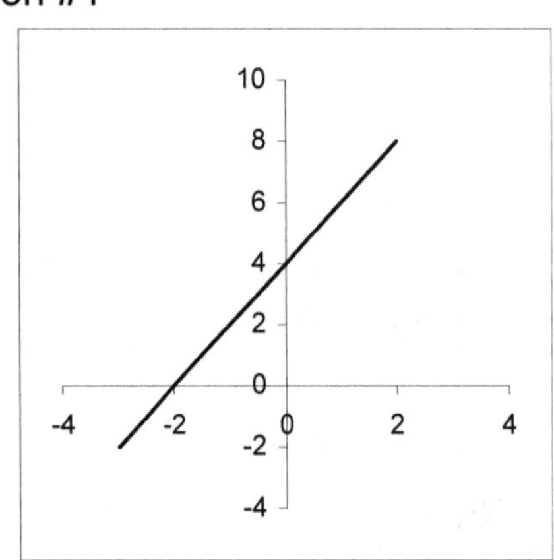

Question #3

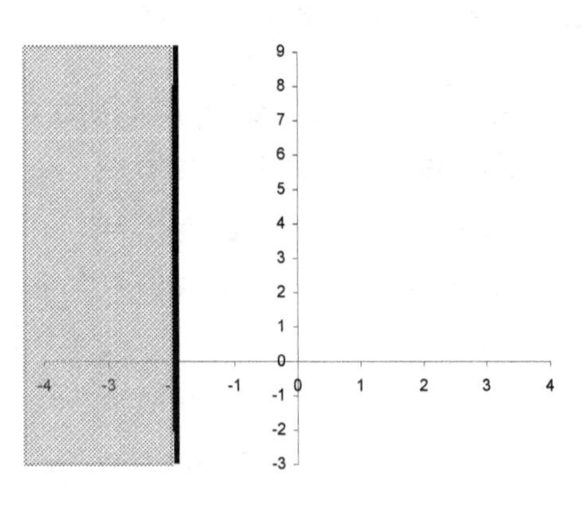

Question #2

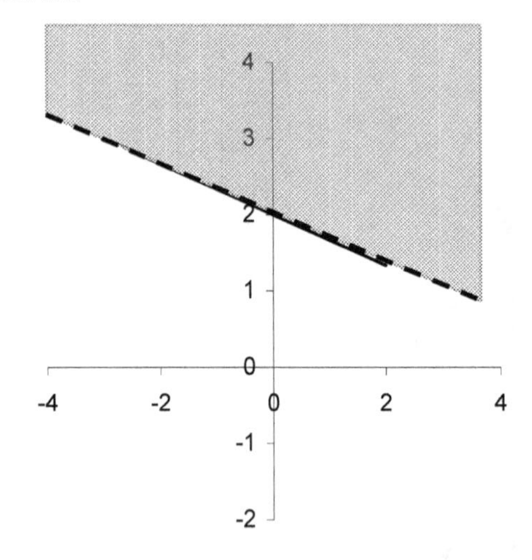

page 29

Question #1 x-intercept = -14 y-intercept = -10 slope = $-\frac{5}{7}$

Question #2 x-intercept = 14 y-intercept = -7 slope = $\frac{1}{2}$

Question #3 x-intercept = 3 y-intercept = none

Question #4 x-intercept = $\frac{15}{2}$ y-intercept = 3 slope = $-\frac{2}{5}$

page 30

Question #1 $y=\frac{3}{4}x+\frac{17}{4}$

Question #2 $x=11$

Question #3 $y=\frac{3}{5}x+\frac{42}{5}$

Question #4 $y=5$

page 40

Question #1

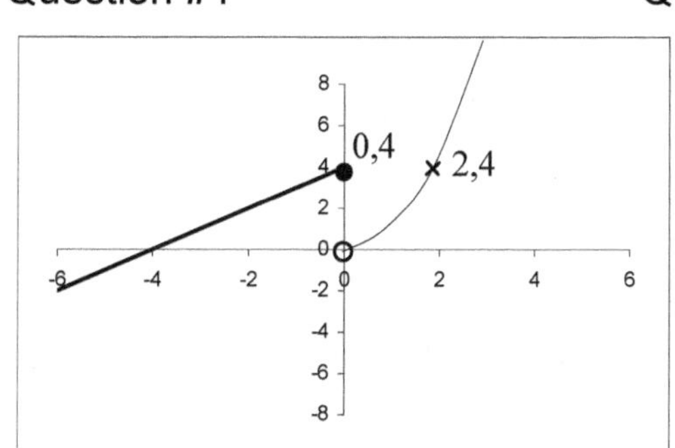

Question #2

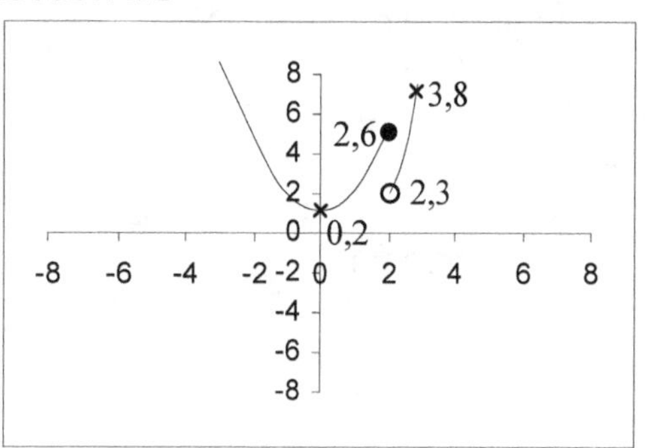

Competency 0009, *page 45*

Question #1 The sides are 8, 15, and 17

Question #2 The numbers are 2 and $\frac{1}{2}$

page 47

Question #1 $x^2 - 10x + 25$

Question #2 $25x^2 - 10x - 48$

Question #3 $x^2 - 9x - 36$

Competency 0010, *page 52*

Question #1 It takes Curly 15 minutes to paint the elephant alone

Question #2 The original fraction is $\frac{1}{3}$.

Question #3 The car was traveling at 68mph and the truck was traveling at 62mph

Competency 0018, *page 130*

Question #1

$$\cot\theta = \frac{x}{y}$$

$$\frac{x}{y} = \frac{x}{r} \times \frac{r}{y} = \frac{x}{y} = \cot\theta$$

Question #2

$$1 + \cot^2\theta = \csc^2\theta$$

$$\frac{y^2}{y^2} + \frac{x^2}{y^2} = \frac{r^2}{y^2} = \csc^2\theta$$

Competency 0020, *page 139*

Question #1 49.34
Question #2 1

page 141

Question #1 ∞
Question #2 -1

Sample Test: Mathematics

1) Given W = whole numbers
N = natural numbers
Z = integers
R = rational numbers
I = irrational numbers

Which of the following is not true?

A) $R \subset I$

B) $W \subset Z$

C) $Z \subset R$

D) $N \subset W$

2) Which of the following is an irrational number?

A) .362626262...

B) $4\frac{1}{3}$

C) $\sqrt{5}$

D) $-\sqrt{16}$

3) Which denotes a complex number?

A) 4.1212121212...

B) $-\sqrt{16}$

C) $\sqrt{127}$

D) $\sqrt{-100}$

4) Choose the correct statement:

A) Rational and irrational numbers are both proper subsets of the real numbers.

B) The set of whole numbers is a proper subset of the set of natural numbers.

C) The set of integers is a proper subset of the set of irrational numbers.

D) The set of real numbers is a proper subset of the natural, whole, integers, rational, and irrational numbers.

5) Which statement is an example of the identity axiom of addition?

A) $3 + -3 = 0$

B) $3x = 3x + 0$

C) $3 \cdot \frac{1}{3} = 1$

D) $3 + 2x = 2x + 3$

6) Which axiom is incorrectly applied?

$3x + 4 = 7$

Step a $3x + 4 - 4 = 7 - 4$

additive equality

Step b $3x + 4 - 4 = 3$

commutative axiom of addition

Step c $3x + 0 = 3$

additive inverse

Step d $3x = 3$

additive identity

A) step a

B) step b

C) step c

D) step d

7) Which of the following sets is closed under division?

A) integers

B) rational numbers

C) natural numbers

D) whole numbers

8) How many real numbers lie between -1 and +l ?

A) 0

B) 1

C) 17

D) an infinite number

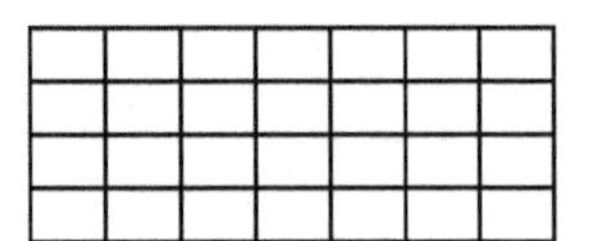

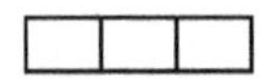

9) The above diagram would be least appropriate for illustrating which of the following?

A) $7 \times 4 + 3$

B) $31 \div 8$

C) 28×3

D) $31 - 3$

10) $24 - 3 \times 7 + 2 =$

A) 5

B) 149

C) -3

D) 189

11) Which of the following does not correctly relate an inverse operation?

A) a - b = a + -b

B) $a \times b = b \div a$

C) $\sqrt{a^2} = a$

D) $a \times \frac{1}{a} = 1$

12) Mr. Brown feeds his cat premium cat food which costs $40 per month. Approximately how much will it cost to feed her for one year?

A) $500

B) $400

C) $80

D) $4800

13) Given that n is a positive even integer, 5n + 4 will always be divisible by:

A) 4

B) 5

C) 5n

D) 2

14) Given that x, y, and z are prime numbers, which of the following is true?

A) x + y is always prime

B) xyz is always prime

C) xy is sometimes prime

D) x + y is sometimes prime

15) Find the GCF of $2^2 \cdot 3^2 \cdot 5$ and $2^2 \cdot 3 \cdot 7$.

A) $2^5 \cdot 3^3 \cdot 5 \cdot 7$

B) $2 \cdot 3 \cdot 5 \cdot 7$

C) $2^2 \cdot 3$

D) $2^3 \cdot 3^2 \cdot 5 \cdot 7$

16) Given even numbers x and y, which could be the LCM of x and y?

A) $\frac{xy}{2}$

B) 2xy

C) 4xy

D) xy

17) $(3.8 \times 10^{17}) \times (.5 \times 10^{-12})$

A) 19×10^5

B) 1.9×10^5

C) 1.9×10^6

D) 1.9×10^7

18) 2^{-3} is equivalent to

A) .8

B) -.8

C) 125

D) 125

19) $\dfrac{3.5 \times 10^{-10}}{0.7 \times 10^4}$

A) 0.5×10^6

B) 5.0×10^{-6}

C) 5.0×10^{-14}

D) 0.5×10^{-14}

20) Solve for x: $\dfrac{4}{x} = \dfrac{8}{3}$

A) .66666...

B) .6

C) 15

D) 1.5

21) Choose the set in which the members are <u>not</u> equivalent.

A) 1/2, 0.5 , 50%

B) 10/5, 2.0 , 200%

C) 3/8, 0.385, 38.5%

D) 7/10, 0.7 , 70%

22) If three cups of concentrate are needed to make 2 gallons of fruit punch, how many cups are needed to make 5 gallons?

A) 6 cups

B) 7 cups

C) 7.5 cups

D) 10 cups

23) A sofa sells for $520. If the retailer makes a 30% profit, what was the wholesale price?

A) $400

B) $676

C) $490

D) $364

24) Given a spinner with the numbers one through eight, what is the probability that you will spin an even number or a number greater than four?

A) 1/4

B) 1/2

C) ¾

D) 1

25) If a horse will probably win three races out of ten, what are the odds that he will win?

A) 3:10

B) 7:10

C) 3:7

D) 7:3

26) Given a drawer with 5 black socks, 3 blue socks, and 2 red socks, what is the probability that you will draw two black socks in two draws in a dark room?

A) 2/9

B) 1/4

C) 17/18

D) 1/18

27) A sack of candy has 3 peppermints, 2 butterscotch drops and 3 cinnamon drops. One candy is drawn and replaced, then another candy is drawn; what is the probability that both will be butterscotch?

A) 1/2

B) 1/28

C) 1/4

D) 1/16

28) Find the median of the following set of data:

14 3 7 6 11 20

A) 9

B) 8.5

C) 7

D) 11

29) Corporate salaries are listed for several employees. Which would be the best measure of central tendency?

$24,000 $24,000 $26,000
$28,000 $30,000 $120,000

A) mean

B.) median

C) mode

D) no difference

30) Which statement is true about George's budget?

A) George spends the greatest portion of his income on food.

B) George spends twice as much on utilities as he does on his mortgage.

C) George spends twice as much on utilities as he does on food.

D) George spends the same amount on food and utilities as he does on mortgage.

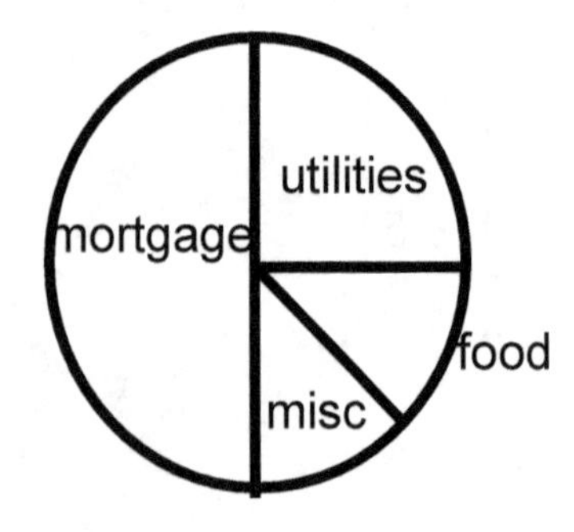

31) A student scored in the 87th percentile on a standardized test. Which would be the best interpretation of his score?

A) Only 13% of the students who took the test scored higher.

B) This student should be getting mostly B's on his report card.

C) This student performed below average on the test.

D) This is the equivalent of missing 13 questions on a 100 question exam.

32) A man's waist measures 90 cm. What is the greatest possible error for the measurement?

A) $\pm$ 1 m

B) $\pm$8 cm

C) $\pm$1 cm

D) $\pm$5 mm

33) The mass of a cookie is closest to

A) 0.5 kg

B) 0.5 grams

C) 15 grams

D) 1.5 grams

34) 3 km is equivalent to

A) 300 cm

B) 300 m

C) 3000 cm

D) 3000 m

35) 4 square yards is equivalent to

A) 12 square feet

B) 48 square feet

C) 36 square feet

D) 108 square feet

36) If a circle has an area of 25 cm2, what is its circumference to the nearest tenth of a centimeter?

A) 78.5 cm

B) 17.7 cm

C) 8.9 cm

D) 15.7 cm

37) Find the area of the figure below.

12 in

3 in →

7 in

5 in

A) 56 in^2

B) 27 in^2

C) 71 in^2

D) 170 in^2

38) Find the area of the shaded region given square ABCD with side AB=10m and circle E.

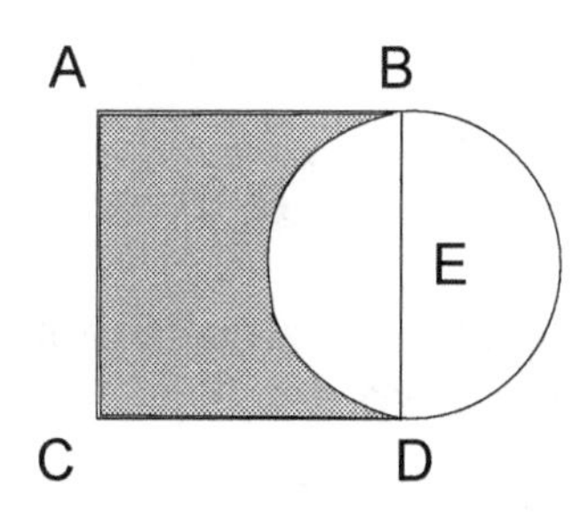

A) 178.5 m^2

B) 139.25 m^2

C) 71 m^2

D) 60.75 m^2

39) Given similar polygons with corresponding sides of lengths 9 and 15, find the perimeter of the smaller polygon if the perimeter of the larger polygon is 150 units.

A) 54

B) 135

C) 90

D) 126

40)

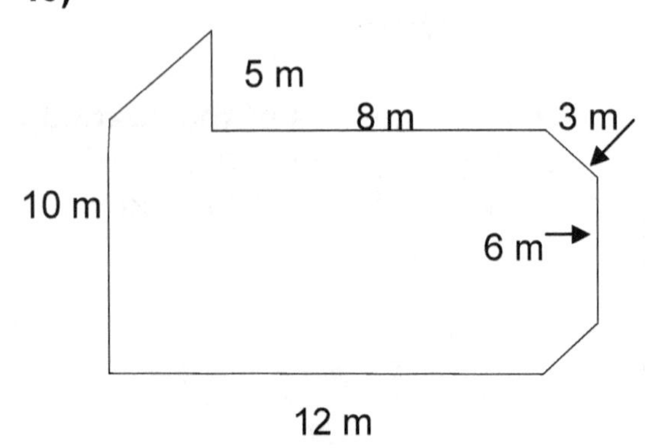

Compute the area of the polygon shown above.

A) 178 m^2

B) 154 m^2

C) 43 m^2

D) 188 m^2

41) If the radius of a right cylinder is doubled, how does its volume change?

A) no change

B) also is doubled

C) four times the original

D) pi times the original

42) Determine the volume of a sphere to the nearest cm^3 if the surface area is 113 cm^2.

A) 113 cm^3

B) 339 cm^3

C) 37.7 cm^3

D) 226 cm3

43) Compute the surface area of the prism.

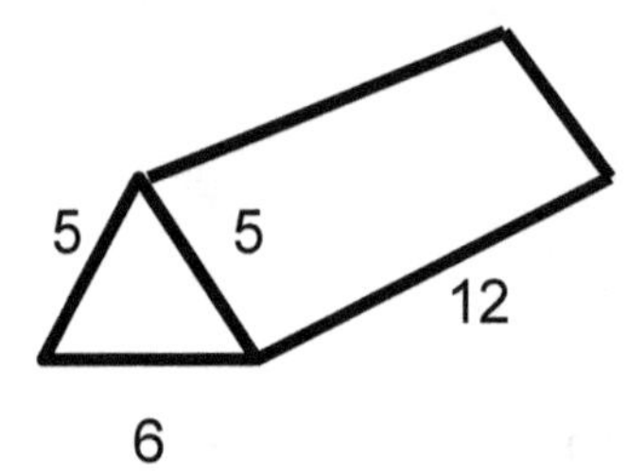

A) 204

B) 216

C) 360

D) 180

44) If the base of a regular square pyramid is tripled, how does its volume change?

A) double the original

B) triple the original

C) nine times the original

D) no change

45) How does lateral area differ from total surface area in prisms, pyramids, and cones?

A) For the lateral area, only use surfaces perpendicular to the base.

B) They are both the same.

C) The lateral area does not include the base.

D) The lateral area is always a factor of pi.

46) Given XY ≅ YZ and ∠AYX ≅ ∠AYZ. Prove ΔAYZ ≅ ΔAYX.

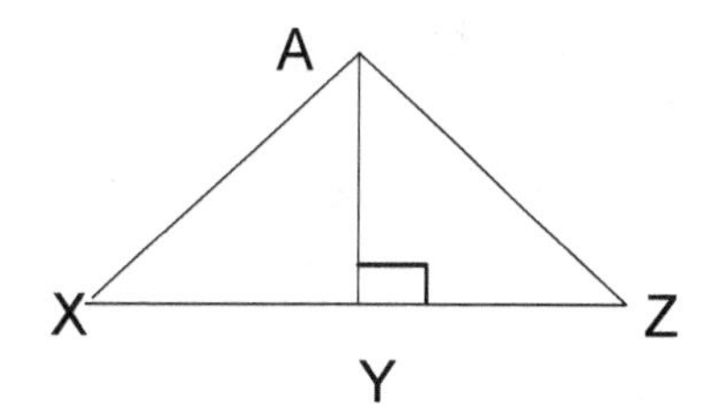

1) XY ≅YZ

2) ∠AYX ≅∠AYZ

3) AY ≅ AY

4) ΔAYZ ≅ ΔAYX

Which property justifies step 3?

A) reflexive

B.) symmetric

B) transitive

D) identity

47) Given $l_1 \parallel l_2$ (parallel lines 1 & 2) prove $\angle b \cong \angle e$

1) $\angle b \cong \angle d$ 1) vertical angle theorem

2) $\angle d \cong \angle e$ 2) alternate interior angle theorem

3) $\angle b \cong \angle 3$ 3) symmetric axiom of equality

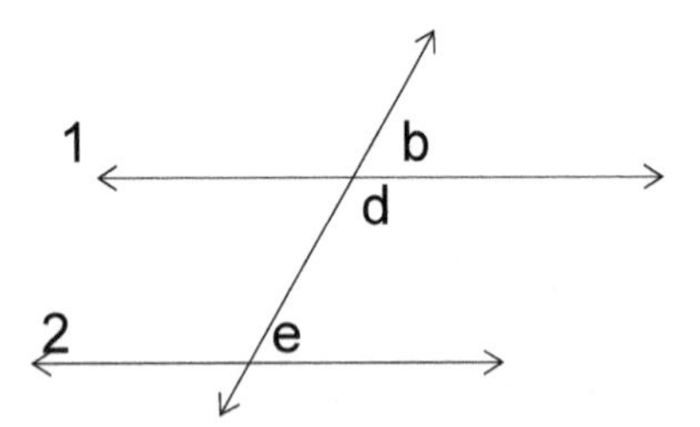

Which step is incorrectly justified?

A) step 1

B) step 2

C) step 3

D) no error

48) Simplify $\dfrac{\frac{3}{4}x^2y^{-3}}{\frac{2}{3}xy}$

A) $\frac{1}{2}xy^{-4}$

B) $\frac{1}{2}x^{-1}y^{-4}$

C) $\frac{9}{8}xy^{-4}$

D) $\frac{9}{8}xy^{-2}$

49) $7t - 4 \cdot 2t + 3t \cdot 4 \div 2 =$

A) 5t

B) 0

C) 31t

D) 18t

50) Solve for x:
$3x + 5 \geq 8 + 7x$

A) $x \geq -\frac{3}{4}$

B) $x \leq -\frac{3}{4}$

C) $x \geq \frac{3}{4}$

D) $x \leq \frac{3}{4}$

51) Solve for x:
$|2x + 3| > 4$

A) $-\frac{7}{2} > x > \frac{1}{2}$

B) $-\frac{1}{2} > x > \frac{7}{2}$

C) $x < \frac{7}{2}$ or $x < -\frac{1}{2}$

D) $x < -\frac{7}{2}$ or $x > \frac{1}{2}$

52) $3x + 2y = 12$
$12x + 8y = 15$

A) all real numbers

B) $x = 4, y = 4$

C) $x = 2, y = -1$

D) $\varnothing$

53) $x = 3y + 7$
$7x + 5y = 23$

A) (-1,4)

B) (4, -1)

C) $(\frac{-29}{7}, \frac{-26}{7})$

D) (10, 1)

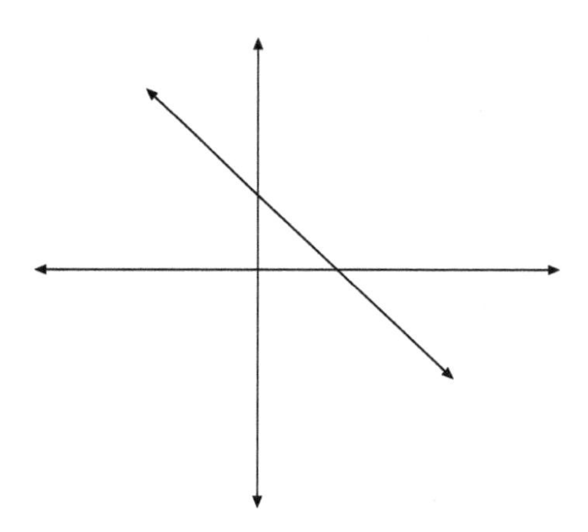

54) Which equation is represented by the above graph?

A) $x - y = 3$

B) $x - y = -3$

C) $x + y = 3$

D) $x + y = -3$

55) Graph the solution:
$|x| + 7 < 13$

A) 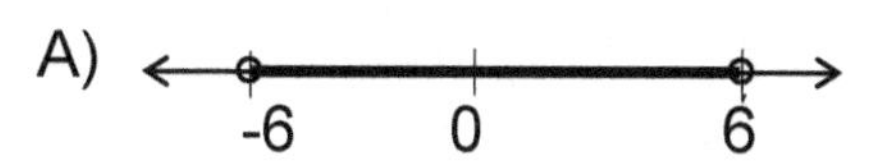

B) -6 0 6

C) -6 0 6

D) 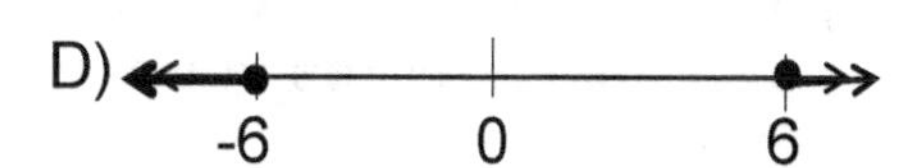

56) Three less than four times a number is five times the sum of that number and 6. Which equation could be used to solve this problem?

A) $3 - 4n = 5(n + 6)$

B) $3 - 4n + 5n = 6$

C) $4n - 3 = 5n + 6$

D) $4n - 3 = 5(n + 6)$

57) A boat travels 30 miles upstream in three hours. It makes the return trip in one and a half hours. What is the speed of the boat in still water?

A) 10 mph

B) 15 mph

C) 20 mph

D) 30 mph

58) Which set illustrates a function?

A) { (0,1) (0,2) (0,3) (0,4) }

B) { (3,9) (-3,9) (4,16) (-4,16)}

C) { (1,2) (2,3) (3,4) (1,4) }

D) { (2,4) (3,6) (4,8) (4,16) }

59) Give the domain for the function over the set of real numbers:

$$y = \frac{3x + 2}{2x - 3}$$

A) all real numbers

B) all real numbers, $x \neq 0$

C) all real numbers, $x \neq -2$ or 3

D) all real numbers, $x \neq \frac{\pm\sqrt{6}}{2}$

60) Factor completely: 8(x - y) + a(y - x)

A) (8 + a) (y - x)

B) (8 - a) (y - x)

C) (a - 8) (y - x)

D) (a - 8) (y + x)

61) Which of the following is a factor of $k^3 - m^3$?

A) $k^2 + m^2$

B) $k + m$

C) $k^2 - m^2$

D) $k - m$

62) Solve for x.

$$3x^2 - 2 + 4(x^2 - 3) = 0$$

A) $\{-\sqrt{2}, \sqrt{2}\}$

B) {2, -2 }

C) $\{0, \sqrt{3}, -\sqrt{3}\}$

D) {7, -7 }

63) Solve: $\sqrt{75} + \sqrt{147} - \sqrt{48}$

A) 174

B) $12\sqrt{3}$

C) $8\sqrt{3}$

D) 74

64) The discriminant of a quadratic equation is evaluated and determined to be -3. The equation has

A) one real root

B) one complex root

C) two roots, both real

D) two roots, both complex

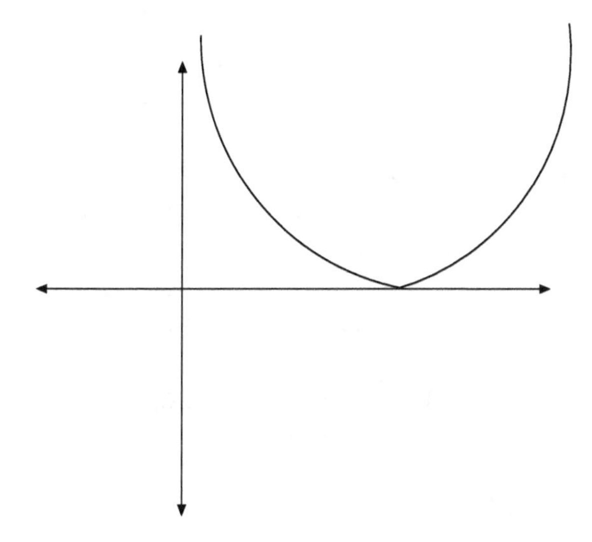

65) Which equation is graphed above?

A) $y = 4(x + 3)^2$

B) $y = 4(x - 3)^2$

C) $y = 3(x - 4)^2$

D) $y = 3(x + 4)^2$

66) If y varies inversely as x and x is 4 when y is 6, what is the constant of variation?

A) 2

B) 12

C) 3/2

D) 24

67) If y varies directly as x and x is 2 when y is 6, what is x when y is 18?

A) 3

B) 6

C) 26

D) 36

68) {1,4,7,10, . . .}

What is the 40th term in this sequence?

A) 43

B) 121

C) 118

D) 120

69) {6,11,16,21, . .}

Find the sum of the first 20 terms in the sequence.

A) 1070

B) 1176

C) 969

D) 1069

70) Two non-coplanar lines which do not intersect are labeled

A) parallel lines

B) perpendicular lines

C) skew lines

D) alternate exterior lines

71)

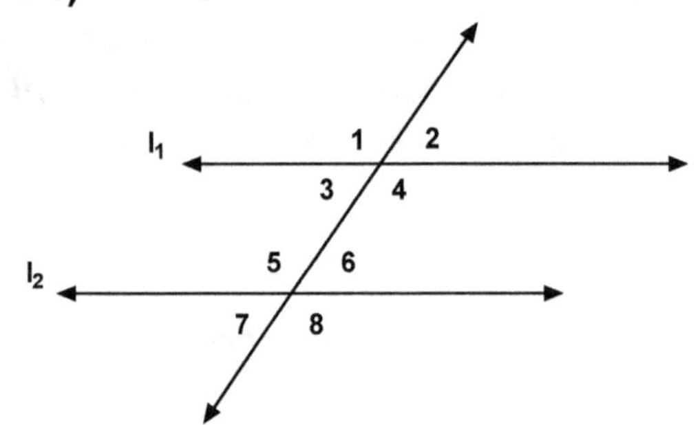

Given $l_1 \parallel l_2$ (parallel lines 1 & 2) which of the following is true?

A) $\angle 1$ and $\angle 8$ are congruent and alternate interior angles

B) $\angle 2$ and $\angle 3$ are congruent and corresponding angles

C) $\angle 3$ and $\angle 4$ are adjacent and supplementary angles

D) $\angle 3$ and $\angle 5$ are adjacent and supplementary angles

72)

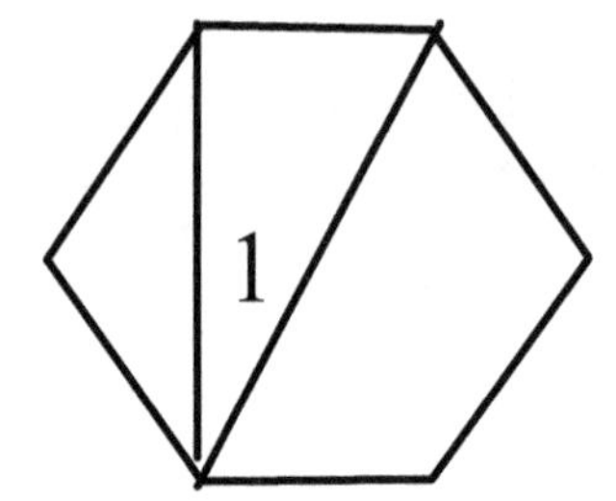

Given the regular hexagon above, determine the measure of angle $\angle 1$.

A) 30°

B) 60°

C) 120°

D) 45°

73)

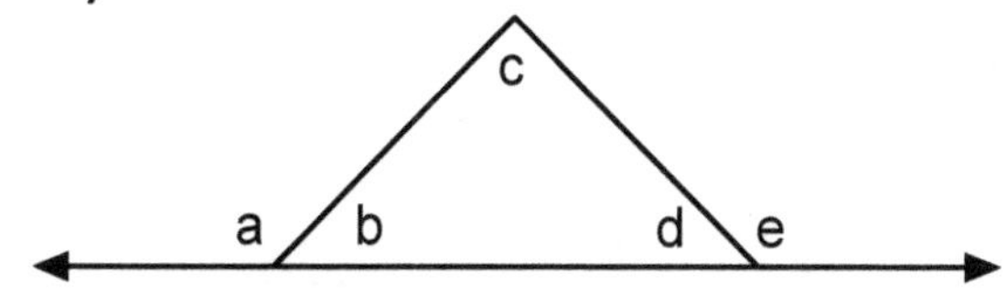

Which of the following statements is true about the number of degrees in each angle?

A) $a + b + c = 180^\circ$

B) $a = e$

C) $b + c = e$

D) $c + d = e$

74)

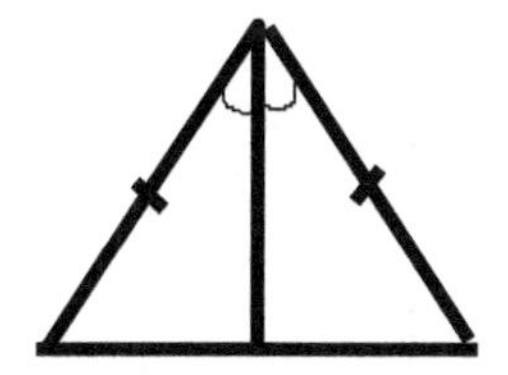

What method could be used to prove the above triangles congruent?

A) SSS

B) SAS

C) AAS

D) SSA

75)

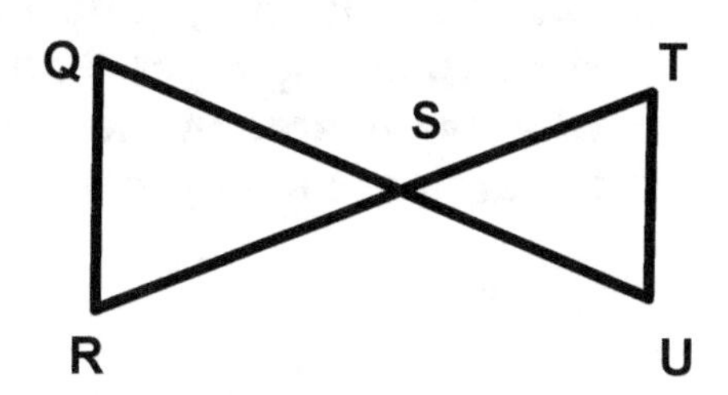

Given $QS \cong TS$ and $RS \cong US$, prove $\Delta QRS \cong \Delta TUS$.

l) $QS \cong TS$	1) Given
2) $RS \cong US$	2) Given
3) $\angle TSU \cong \angle QSR$	3) ?
4) $\Delta TSU \cong \Delta QSR$	4) SAS

Give the reason which justifies step 3.

A) Congruent parts of congruent triangles are congruent

B) Reflexive axiom of equality

C) Alternate interior angle Theorem

D) Vertical angle theorem

76) Given similar polygons with corresponding sides 6 and 8, what is the area of the smaller if the area of the larger is 64?

A) 48

B) 36

C) 144

D) 78

77) In similar polygons, if the perimeters are in a ratio of x:y, the sides are in a ratio of

A) x:y

B) x^2:y^2

C) 2x:y

D) 1/2 x:y

78)

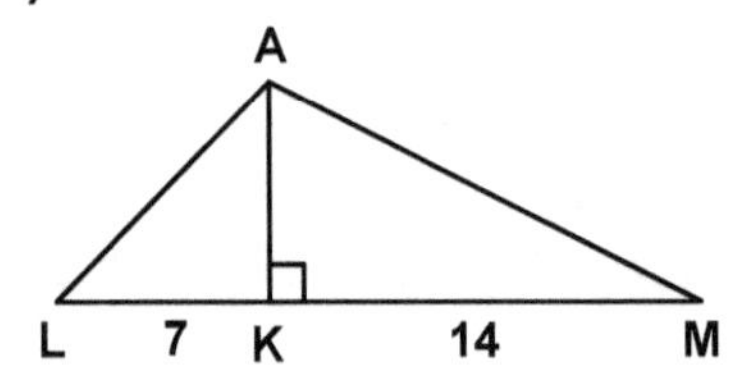

Given altitude AK with measurements as indicated, determine the length of AK.

A) 98

B) $7\sqrt{2}$

C) $\sqrt{21}$

D) $7\sqrt{3}$

79)

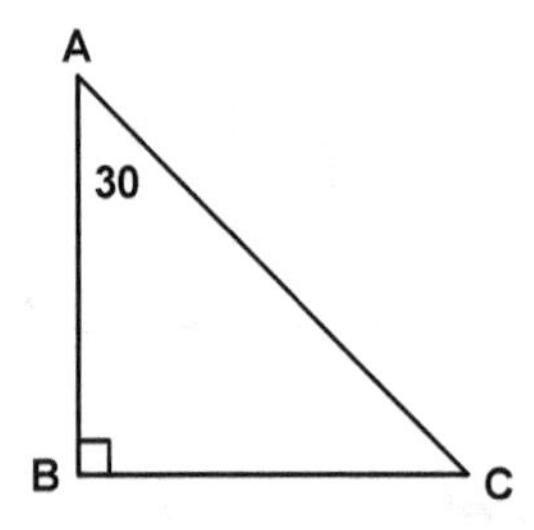

If AC = 12, determine BC.

A) 6

B) 4

C) $6\sqrt{3}$

D) $3\sqrt{6}$

80)

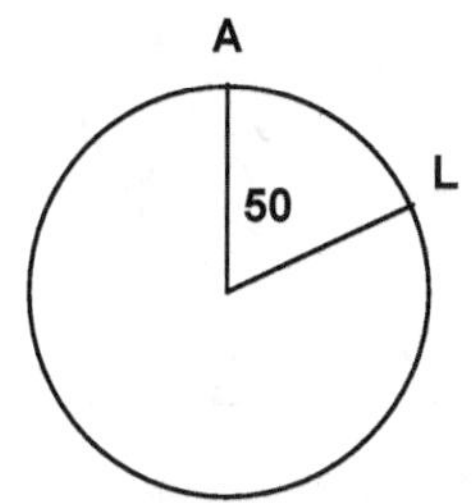

What is the measure of major arc AL ?

A) 50°

B) 25°

C) 100°

D) 310°

81)

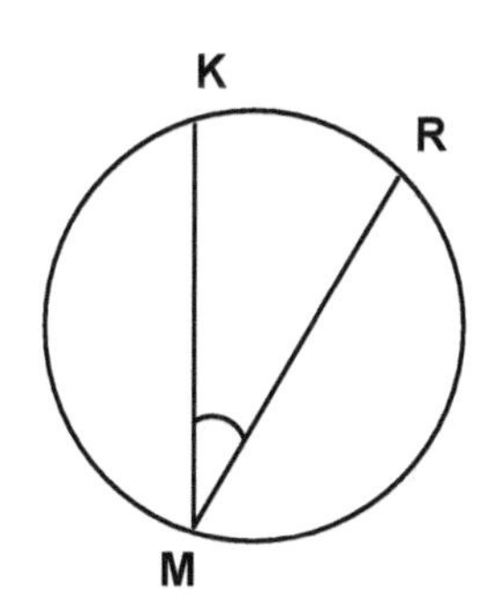

If arc KR = 70° what is the measure of ∠M?

A) 290°

B) 35°

C) 140°

D) 110°

82)

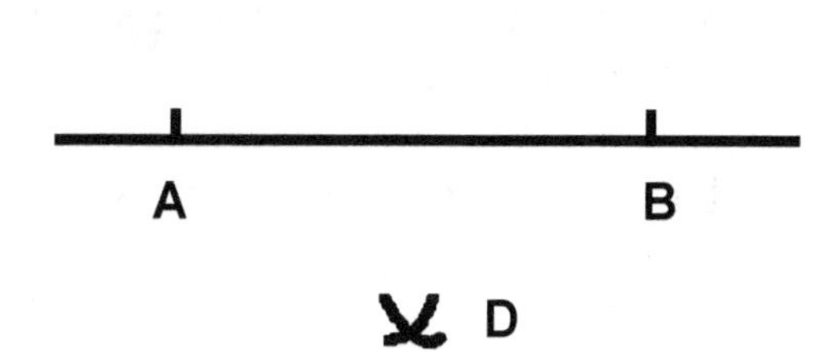

The above construction can be completed to make:

A) an angle bisector

B) parallel lines

C) a perpendicular bisector

D) skew lines

83)

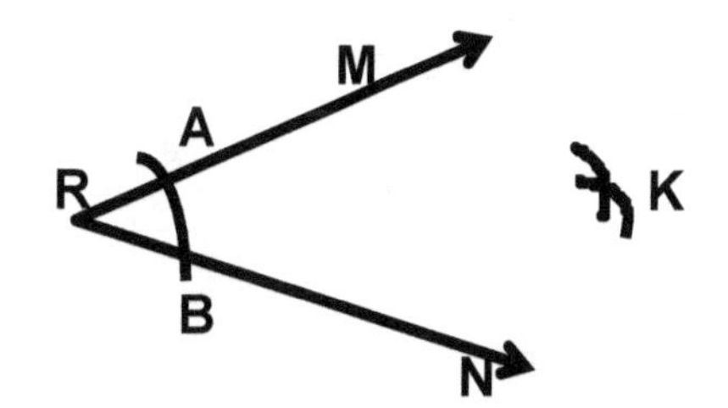

A line from R to K will form

A) an altitude of RMN

B) a perpendicular bisector of MN

C) a bisector of MRN

D) a vertical angle

84) Which is a postulate?

A) The sum of the angles in any triangle is 180°.

B) A line intersects a plane in one point.

C) Two intersecting lines from congruent vertical angles.

D) Any segment is congruent to itself.

85) Which of the following can be defined?

A) point

B) ray

C) line

D) plane

86)

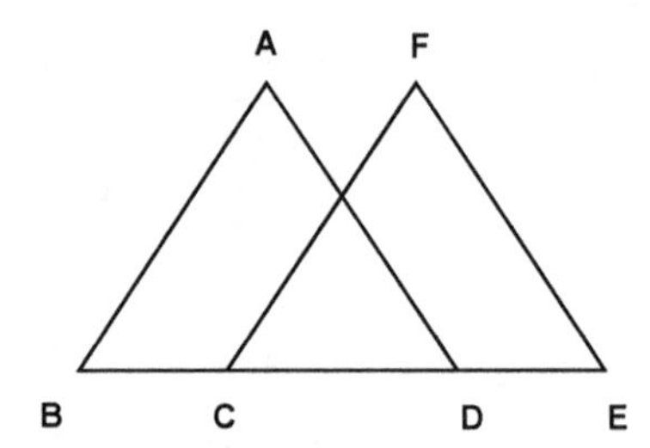

Which theorem could be used to prove $\Delta ABD \cong \Delta CEF$, given $BC \cong DE$, $\angle C \cong \angle D$, and $AD \cong CF$?

A) ASA

B) SAS

C) SAA

D) SSS

87)

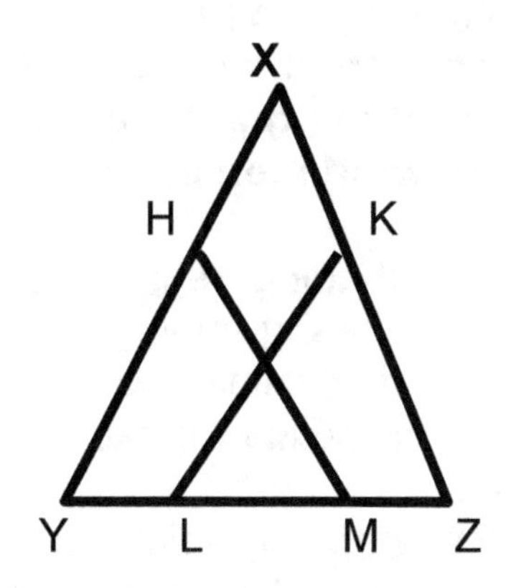

Prove $\Delta HYM \cong \Delta KZL$, given $XZ \cong XY$, $\angle L \cong \angle M$ and $YL \cong MZ$

1) $XZ \cong XY$	1) Given
2) $\angle Y \cong \angle Z$	2) ?
3) $\angle L \cong \angle M$	3) Given
4) $YL \cong MZ$	4) Given
5) $LM \cong LM$	5) ?
6) $YM \cong LZ$	6) Add
7) $\Delta HYM \cong \Delta KZL$	7) ASA

Which could be used to justify steps 2 and 5?

A) CPCTC, Identity

B) Isosceles Triangle Theorem, Identity

C) SAS, Reflexive

D) Isosceles Triangle Theorem, Reflexive

88) Find the distance between (3,7) and (-3,4).

A) 9

B) 45

C) $3\sqrt{5}$

D) $5\sqrt{3}$

89) Find the midpoint of (2,5) and (7,-4).

A) (9,-1)

B) (5,9)

C) (9/2 , -1/2)

D) (9/2, 1/2)

90) Given segment AC with B as its midpoint find the coordinates of C if A = (5,7) and B = (3, 6.5).

A) (4, 6.5)

B) (1, 6)

C) (2, 0.5)

D) (16, 1)

91)

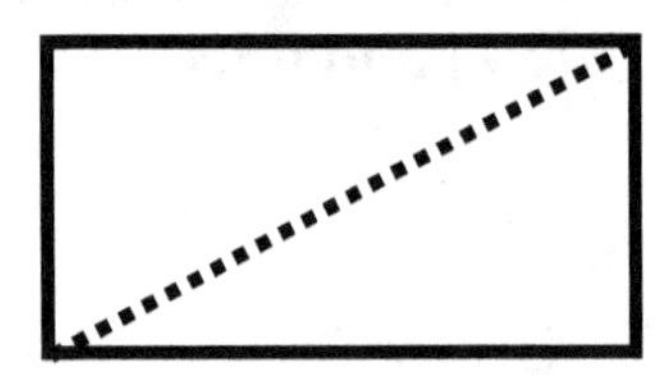

The above diagram is most likely used in deriving a formula for which of the following?

A) the area of a rectangle

B) the area of a triangle

C) the perimeter of a triangle

D) the surface area of a prism

92) **A student turns in a paper with this type of error:**

$7 + 16 \div 8 \times 2 = 8$
$8 - 3 \times 3 + 4 = -5$

In order to remediate this error, a teacher should:

A) review and drill basic number facts

B) emphasize the importance of using parentheses in simplifying expressions

C) emphasize the importance of working from left to right when applying the order of operations

D) do nothing; these answers are correct

93) **Identify the proper sequencing of subskills when teaching graphing inequalities in two dimensions**

A) shading regions, graphing lines, graphing points, determining whether a line is solid or broken

B) graphing points, graphing lines, determining whether a line is solid or broken, shading regions

C) graphing points, shading regions, determining whether a line is solid or broken, graphing lines

D) graphing lines, determining whether a line is solid or broken, graphing points, shading regions

94) **Sandra has $34.00, Carl has $42.00. How much more does Carl have than Sandra?**

Which would be the best method for finding the answer?

A) addition

B) subtraction

C) division

D) both A and B are equally correct

95) Which is the least appropriate strategy to emphasize when teaching problem solving?

A) guess and check

B) look for key words to indicate operations such as all together-add, more than-subtract, times-multiply

C) make a diagram

D) solve a simpler version of the problem

96) Choose the least appropriate set of manipulatives for a six grade class.

A) graphic calculators, compasses, rulers, conic section models

B) two color counters, origami paper, markers, yarn

C) balance, meter stick, colored pencils, beads

D) paper cups, beans, tangrams, geoboards

97) According to Piaget, at which developmental level would a child be able to learn formal algebra?

A) pre-operational

B) sensory-motor

C) abstract

D) concrete operational

98) Which statement is incorrect?

A) Drill and practice is one good use for classroom computers.

B) Some computer programs can help to teach problem solving.

C) Computers are not effective unless each child in the class has his own workstation.

D) Analyzing science project data on a computer during math class is an excellent use of class time.

98) Given a,b,y, and z are real numbers and ay + b = z, Prove

$$y = \frac{z + -b}{a}$$

Statement	Reason
1) ay + b = z	1) Given
2) -b is a real number	2) Closure
3) (ay +b) + -b = z + -b	3) Addition property of Identity
4) ay + (b + -b) = z + -b	4) Associative
5) ay + 0 = z + -b	5) Additive inverse
6) ay = z + -b	6) Addition property of identity
7) $a = \frac{z + -b}{y}$	7) Division

99) Which reason is incorrect for the corresponding statement?

A) step 3

B) step 4

C) step 5

D) step 6

100) Seventh grade students are working on a project using non-standard measurement. Which would not be an appropriate instrument for measuring the length of the classroom?

A) a student's foot

B) a student's arm span

C) a student's jump

D) all are appropriate

101. Change $.\overline{63}$ into a fraction in simplest form.

A) 63/100
B) 7/11
C) 6 3/10
D) 2/3

102. Which of the following sets is closed under division?

I) {½, 1, 2, 4}
II) {-1, 1}
III) {-1, 0, 1}

A) I only
B) II only
C) III only
D) I and II

103. Which of the following illustrates an inverse property?

A) a + b = a - b
B) a + b = b + a
C) a + 0 = a
D) a + (-a) =0

104. $f(x) = 3x - 2;\ f^{-1}(x) =$

A) $3x + 2$
B) $x/6$
C) $2x - 3$
D) $(x + 2)/3$

105. What would be the total cost of a suit for $295.99 and a pair of shoes for $69.95 including 6.5% sales tax?

A) $389.73
B) $398.37
C) $237.86
D) $315.23

106. A student had 60 days to appeal the results of an exam. If the results were received on March 23, what was the last day that the student could appeal?

A) May 21
B) May 22
C) May 23
D) May 24

107. Which of the following is always composite if *x* is odd, *y* is even, and both *x* and *y* are greater than or equal to 2?

A) $x+y$
B) $3x+2y$
C) $5xy$
D) $5x+3y$

108. Which of the following is incorrect?

A) $(x^2y^3)^2 = x^4y^6$
B) $m^2(2n)^3 = 8m^2n^3$
C) $(m^3n^4)/(m^2n^2) = mn^2$
D) $(x+y^2)^2 = x^2+y^4$

109. Express .0000456 in scientific notation.

A) $4.56x10^{-4}$
B) $45.6x10^{-6}$
C) $4.56x10^{-6}$
D) $4.56x10^{-5}$

110. Compute the area of the shaded region, given a radius of 5 meters. 0 is the center.

A) 7.13 cm²
B) 7.13 m²
C) 78.5 m²
D) 19.63 m²

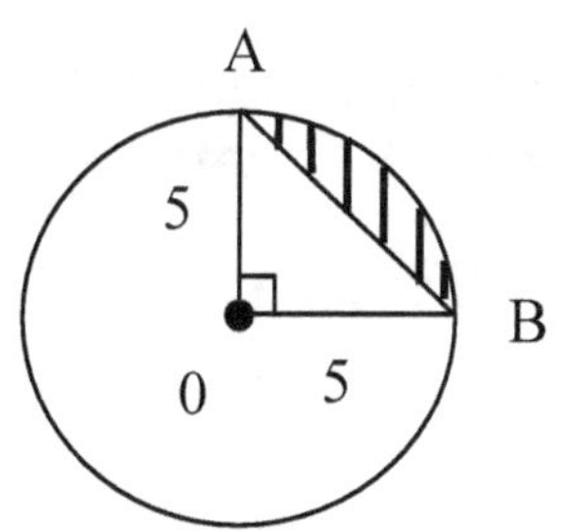

111. If the area of the base of a cone is tripled, the volume will be

A) the same as the original
B) 9 times the original
C) 3 times the original
D) 3 π times the original

112. Find the area of the figure pictured below.

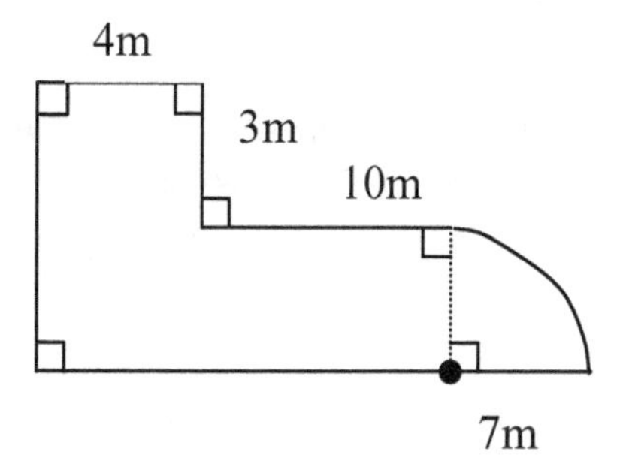

A) 136.47 m²
B) 148.48 m²
C) 293.86 m²
D) 178.47 m²

113. The mass of a Chips Ahoy cookie would be approximately equal to:

A) 1 kilogram
B) 1 gram
C) 15 grams
D) 15 milligrams

114. Compute the median for the following data set:

{12, 19, 13, 16, 17, 14}

A) 14.5
B) 15.17
C) 15
D) 16

115. Half the students in a class scored 80% on an exam, most of the rest scored 85% except for one student who scored 10%. Which would be the best measure of central tendency for the test scores?

A) mean
B) median
C) mode
D) either the median or the mode because they are equal

116. What conclusion can be drawn from the graph below?

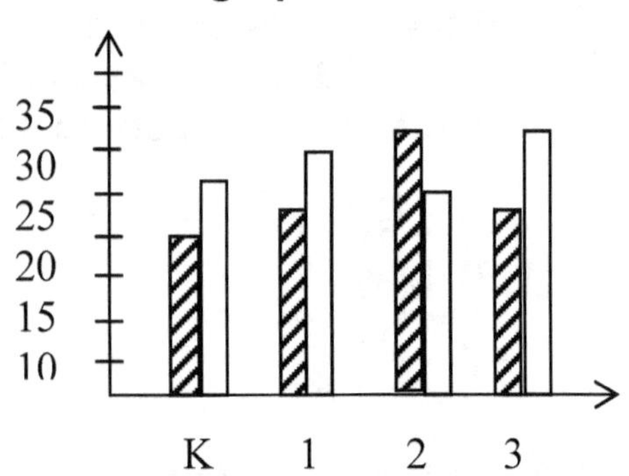

MLK Elementary Student Enrollment ▨ **Girls** □ **Boys**

A) The number of students in first grade exceeds the number in second grade.
B) There are more boys than girls in the entire school.
C) There are more girls than boys in the first grade.
D) Third grade has the largest number of students.

117) State the domain of the function $f(x)=\dfrac{3x-6}{x^2-25}$

A) $x\neq 2$
B) $x\neq 5,-5$
C) $x\neq 2,-2$
D) $x\neq 5$

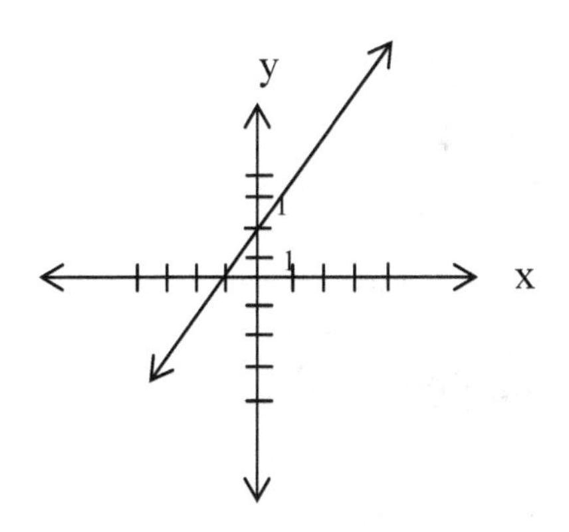

118. What is the equation of the above graph?

A) $2x+y=2$
B) $2x-y=-2$
C) $2x-y=2$
D) $2x+y=-2$

119. Solve for v_0 : $d=at(v_t-v_0)$

A) $v_0=atd-v_t$
B) $v_0=d-atv_t$
C) $v_0=atv_t-d$
D) $v_0=(atv_t-d)/at$

120. Which of the following is a factor of $6+48m^3$

A) (1 + 2m)
B) (1 - 8m)
C) (1 + m - 2m)
D) (1 - m + 2m)

121. Which graph represents the equation of $y=x^2+3x$**?**

A) B)

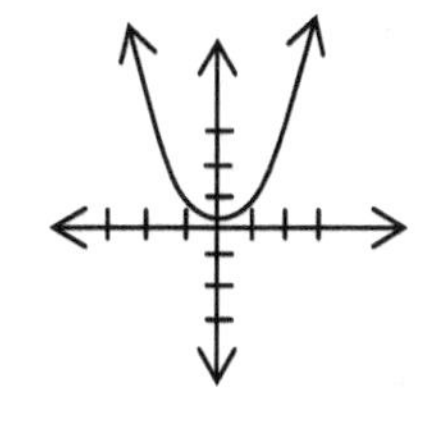
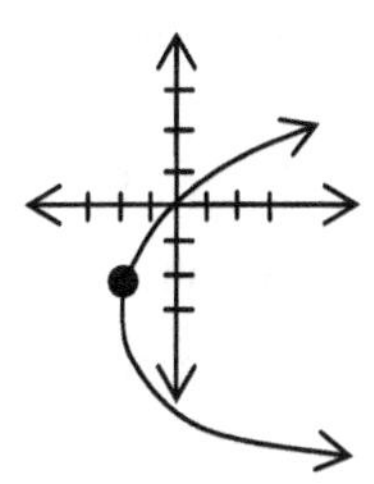

C) D)

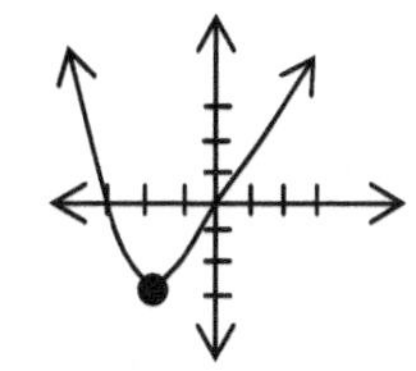
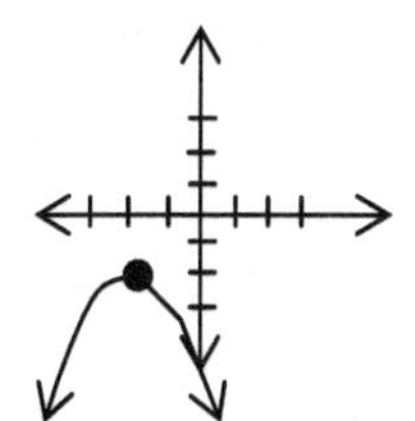

122. The volume of water flowing through a pipe varies directly with the square of the radius of the pipe. If the water flows at a rate of 80 liters per minute through a pipe with a radius of 4 cm, at what rate would water flow through a pipe with a radius of 3 cm?

A) 45 liters per minute
B) 6.67 liters per minute
C) 60 liters per minute
D) 4.5 liters per minute

123) Solve the system of equations for x, y and z.

$$3x + 2y - z = 0$$
$$2x + 5y = 8z$$
$$x + 3y + 2z = 7$$

A) $(-1,\ 2,\ 1)$
B) $(1,\ 2,\ -1)$
C) $(-3,\ 4,\ -1)$
D) $(0,\ 1,\ 2)$

124. Solve for x: $18 = 4 + |2x|$

A) $\{-11, 7\}$
B) $\{-7, 0, 7\}$
C) $\{-7, 7\}$
D) $\{-11, 11\}$

125. Which graph represents the solution set for $x^2 - 5x > -6$?

A) -2 0 2

B) -3 0 3

C) -2 0 2

D) -3 0 2 3

126. Find the zeroes of $f(x) = x^3 + x^2 - 14x - 24$

A) 4, 3, 2
B) 3, -8
C) 7, -2, -1
D) 4, -3, -2

127. Evaluate $3^{1/2}(9^{1/3})$

A) $27^{5/6}$
B) $9^{7/12}$
C) $3^{5/6}$
D) $3^{6/7}$

128. Simplify: $\sqrt{27} + \sqrt{75}$

A) $8\sqrt{3}$
B) 34
C) $34\sqrt{3}$
D) $15\sqrt{3}$

129. Simplify: $\frac{10}{1+3i}$

A) $-1.25(1-3i)$
B) $1.25(1+3i)$
C) $1+3i$
D) $1-3i$

130. Find the sum of the first one hundred terms in the progression. (-6, -2, 2 . . .)

A) 19,200
B) 19,400
C) -604
D) 604

131. How many ways are there to choose a potato and two green vegetables from a choice of three potatoes and seven green vegetables?

A) 126
B) 63
C) 21
D) 252

132. What would be the seventh term of the expanded binomial $(2a+b)^8$?

A) $2ab^7$
B) $41a^4b^4$
C) $112a^2b^6$
D) $16ab^7$

133. Which term most accurately describes two coplanar lines without any common points?

A) perpendicular
B) parallel
C) intersecting
D) skew

134. Determine the number of subsets of set *K*.
***K* = {4, 5, 6, 7}**

A) 15
B) 16
C) 17
D) 18

135. What is the ~~degree~~ measure of each interior angle of a regular 10 sided polygon?

A) 18°
B) 36°
C) 144°
D) 54°

136. If a ship sails due south 6 miles, then due west 8 miles, how far is it from its starting point?

A) 100 miles
B) 10 miles
C) 14 miles
D) 48 miles

137. What is the measure of minor arc AD, given measure of arc PS is 40° and $m<K=10^\circ$?

A) 50°
B) 20°
C) 30°
D) 25°

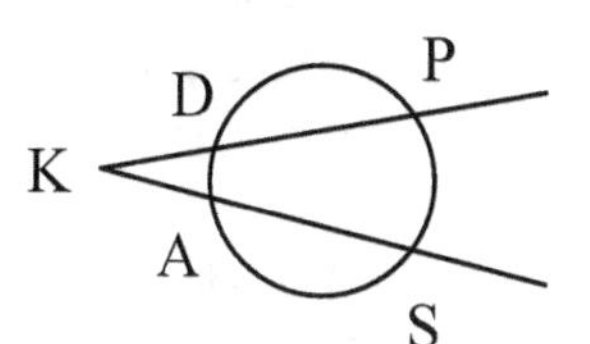

138. Choose the diagram which illustrates the construction of a perpendicular to the line at a given point on the line.

A)

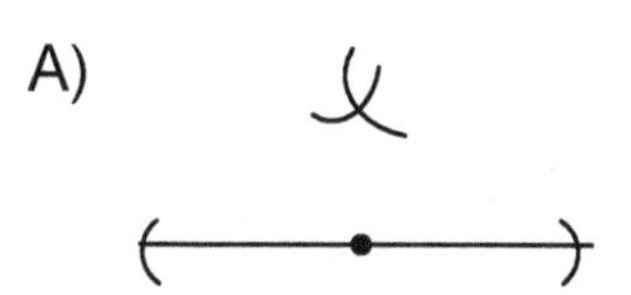

B)

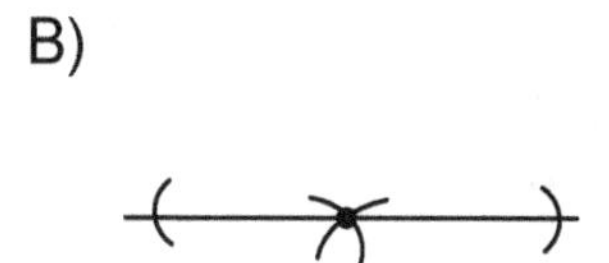

C)

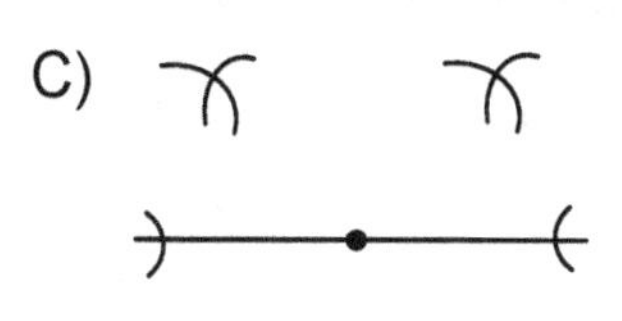

D)

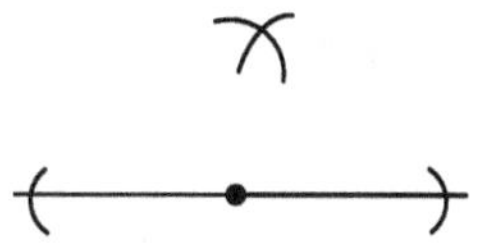

139. When you begin by assuming the conclusion of a theorem is false, then show that through a sequence of logically correct steps you contradict an accepted fact, this is known as

A) inductive reasoning
B) direct proof
C) indirect proof
D) exhaustive proof

140. Which theorem can be used to prove $\Delta BAK \cong \Delta MKA$?

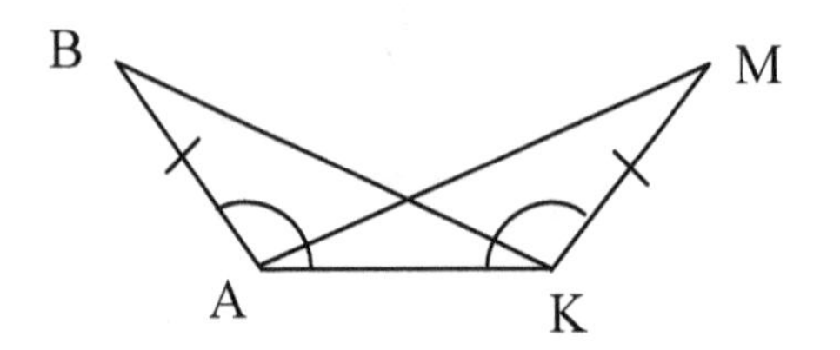

A) SSS
B) ASA
C) SAS
D) AAS

141. Given that QO⊥NP and QO=NP, quadrilateral NOPQ can most accurately be described as a

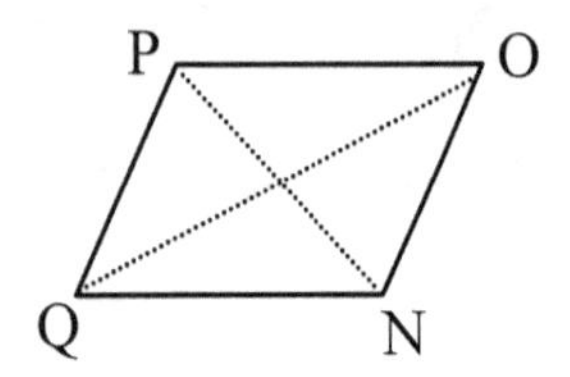

A) parallelogram
B) rectangle
C) square
D) rhombus

142. Choose the correct statement concerning the median and altitude in a triangle.

A) The median and altitude of a triangle may be the same segment.
B) The median and altitude of a triangle are always different segments.
C) The median and altitude of a right triangle are always the same segment.
D) The median and altitude of an isosceles triangle are always the same segment.

143. Which mathematician is best known for his work in developing non-Euclidean geometry?

A) Descartes
B) Riemann
C) Pascal
D) Pythagoras

144. Find the surface area of a box which is 3 feet wide, 5 feet tall, and 4 feet deep.

A) 47 sq. ft.
B) 60 sq. ft.
C) 94 sq. ft
D) 188 sq. ft.

145. Given a 30 meter x 60 meter garden with a circular fountain with a 5 meter radius, calculate the area of the portion of the garden not occupied by the fountain.

A) 1721 m²
B) 1879 m²
C) 2585 m²
D) 1015 m²

146. Determine the area of the shaded region of the trapezoid in terms of *x* and *y*.

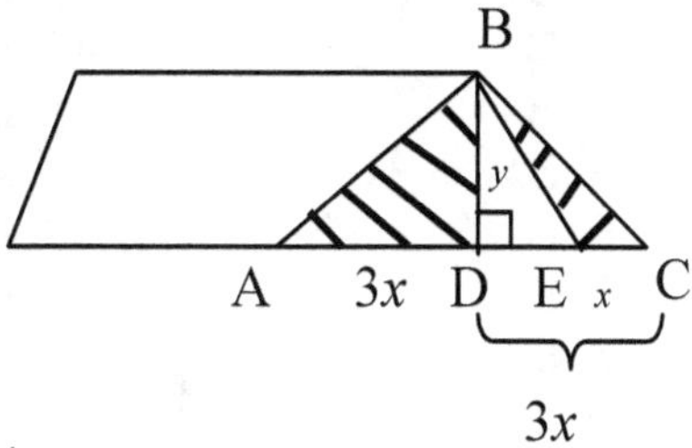

A) $4xy$
B) $2xy$
C) $3x^2y$
D) There is not enough information given.

Answer Key: Mathematics

1. A
2. C
3. D
4. A
5. B
6. B
7. B
8. D
9. C
10. A
11. B
12. A
13. D
14. D
15. C
16. A
17. B
18. D
19. C
20. D
21. C
22. C
23. A
24. C
25. C
26. A
27. D
28. A
29. B
30. C
31. A
32. C
33. C
34. D
35. C
36. B
37. A
38. D
39. C
40. B
41. C
42. A
43. B
44. B
45. C
46. A
47. C
48. C
49. A
50. B
51. D
52. D
53. B
54. C
55. A
56. D
57. B
58. B
59. D
60. C
61. D
62. A
63. C
64. D
65. B
66. D
67. B
68. C
69. A
70. C
71. C
72. A
73. C
74. C
75. D
76. B
77. A
78. B
79. A
80. D
81 B
82. C
83. C
84. D
85. B
86. B
87. D
88. C
89. D
90. B
91. 2
92. C
93. B
94. D
95. B
96. A
97. C
98. C
99. A
100.D
101.B
102.B
103.D
104.D
105.A
106.B
107.C
108.D
109.D
110.B
111.C
112.B
113.C
114.C
115.B
116.B
117.B
118.B
119.D
120.A
121.C
122.A
123.A
124.C
125.D
126.D
127.B
128.A
129.D
130.A
131.A
132.C
133.B
134.B
135.C
136.B
137.B
138.D
139.C
140.C
141.C
142.A
143.B
144.C
145.A
146.B

Rationales for Sample Questions: Mathematics

The following statements represent one way to solve each problem and obtain a correct answer.

1) C The rational numbers are not a subset of the irrational numbers. All of the other statements are true.

2) C 5 is an irrational number. A and B can both be expressed as fractions. D can be simplified to -4, an integer and rational number.

3) D A complex number is the square root of a negative number. The complex number is defined as the square root of -1. A is rational, B and C are irrational.

4) A A proper subset is completely contained in, but not equal to, the original set.

5) B Illustrates the identity axiom of addition. A illustrates additive inverse, C illustrates the multiplicative inverse, and D illustrates the commutative axiom of addition.

6) B In simplifying from step a to step b, 3 replaced 7 - 4, therefore the correct justification would be subtraction or substitution.

7) B In order to be closed under division, when any two members of the set are divided the answer must be contained in the set. This is not true for integers, natural, or whole numbers as illustrated by the counter example 11/2 = 5.5.

8) D There are an infinite number of real numbers between any two real numbers.

9) C is inappropriate. A shows a 7x4 rectangle with 3 additional units. B is the division based on A . D shows how mental subtraction might be visualized leaving a composite difference.

10) A According to the order of operations, multiplication is performed first, then addition and subtraction from left to right.

11) B is always false. A, C, and D illustrate various properties of inverse relations.

12) A 12(40) = 480 which is closest to $500.

13) D 5n is always even. An even number added to an even number is always an even number, thus divisible by 2.

14) D x + y is sometimes prime. B and C show the products of two numbers which are always composite. x + y may be true, but not always,

15) C Choose the number of each prime factor that is in common.

16) A Although choices B, C and D are common multiples, when both numbers are even, the product can be divided by two to obtain the least common multiple.

17) B Multiply the decimals and add the exponents.

18) D Express as the fraction 1/8, then convert to a decimal.

19) C Divide the decimals and subtract the exponents.

20) D Cross multiply to obtain 12 = 8x, then divide both sides by 8.

21) C 3/8 is equivalent to .375 and 37.5%

22) C Set up the proportion 3/2 = x/5, cross multiply to obtain 15=2x, then divide both sides by 2.

23) A Let x be the wholesale price, then x + .30x = 520, 1.30x = 520. Divide both sides by 1.30.

24) C There are 8 favorable outcomes: 2,4,5,6,7,8 and 8 possibilities. Reduce 6/8 to 3/4.

25) C The odds are that he will win 3 and lose 7.

26) A In this example of conditional probability, the probability of drawing a black sock on the first draw is 5/10. It is implied in the problem that there is no replacement, therefore the probability of obtaining a black sock in the second draw is 4/9. Multiply the two probabilities and reduce to lowest terms.

27) D With replacement, the probability of obtaining a butterscotch on the first draw is 2/8 and the probability of drawing a butterscotch on the second draw is also 2/8. Multiply and reduce to lowest terms.

28) A Place the numbers in ascending order: 3 6 7 11 14 20. Find the average of the middle two numbers (7+11)12 =9

29) B The median provides the best measure of central tendency in this case, where the mode is the lowest number and the mean would be disproportionately skewed by the outlier $120,000.

30) C George spends twice as much on utilities as on food.

31) A Percentile ranking tells how the student compared to the norm or the other students taking the test. It does not correspond to the percentage answered correctly, but can indicate how the student compared to the average student tested.

32) D The greatest possible error of measurement is $\pm$+ 1/2 unit, in this case .5 cm or 5 mm.

33) C A cookie is measured in grams.

34) D To change kilometers to meters, move the decimal 3 places to the right.

35) C There are 9 square feet in a square yard.

36) B Find the radius by solving $\Pi r^2 = 25$. Then substitute r=2.82 into $C = 2\Pi r$ to obtain the circumference.

37) A Divide the figure into two rectangles with a horizontal line. The area of the top rectangle is 36 in, and the bottom is 20 in.

38) D Find the area of the square $10^2 = 100$, then subtract 1/2 the area of the circle. The area of the circle is $\Pi r^2 = (3.14)(5)(5)=78.5$. Therefore the area of the shaded region is 100 - 39.25 - 60.75.

39) C The perimeters of similar polygons are directly proportional to the lengths of their sides, therefore 9/15 = x/150. Cross multiply to obtain 1350 = 15x, then divide by 15 to obtain the perimeter of the smaller polygon.

40) B Divide the figure into a triangle, a rectangle and a trapezoid. The area of the triangle is 1/2 bh = 1/2 (4)(5) = 10. The area of the rectangle is bh = 12(10) = 120. The area of the trapezoid is 1/2(b + B)h = 1/2(6 + 10)(3) = 1/2 (16)(3) = S4. Thus, the area of the figure is 10 + 120 + 24 =154.

41) C If the radius of a right cylinder is doubled, the volume is multiplied by four; because in the formula, the radius is squared. Therefore the new volume is 2 x 2 or four times the original.

42) A Solve for the radius of the sphere using $A = 4\Pi r^2$. The radius is 3. Then, find the volume using $4/3\ \Pi r^3$. Only when the radius is 3 are the volume and surface area equivalent.

43) B There are five surfaces which make up the prism. The bottom rectangle has an area 6 x 12 = 72. The sloping sides are two rectangles each with an area of 5 x 12 = 60. The height of the end triangles is determined to be 4 using the Pythagorean theorem. Therefore each triangle has area 1/2bh = 1/2(6)(4) -12. Thus, the surface area is 72 + 60 + 60 + 12 + 12 = 216.

44) B Using the general formula for a pyramid V = 1/3 bh, since the base is tripled and is not squared or cubed in the formula, the volume is also tripled.

45) C The lateral area does not include the base.

46) A The reflexive property states that every number or variable is equal to itself and every segment is congruent to itself.

47) C Step 3 can be justified by the transitive property.

48) C Simplify the complex fraction by inverting the denominator and multiplying: 3/4(3/2)=9/8, then subtract exponents to obtain the correct answer.

49) A First perform multiplication and division from left to right; 7t -8t + 6t, then add and subtract from left to right.

50) B Using additive equality, $-3 \geq 4x$. Divide both sides by 4 to obtain $-3/4 \geq x$. Carefully determine which answer choice is equivalent.

51) D The quantity within the absolute value symbols must be either > 4 or < -4. Solve the two inequalities $2x + 3 > 4$ or $2x + 3 < -4$

52) D Multiplying the top equation by -4 and adding results in the equation 0 = -33. Since this is a false statement, the correct choice is the null set.

53) B Substituting x in the second equation results in 7(3y + 7) + 5y = 23. Solve by distributing and grouping like terms: 26y+49 = 23, 26y = -26, y = -1 Substitute y into the first equation to obtain x.

54) C By looking at the graph, we can determine the slope to be -1 and the y-intercept to be 3. Write the slope intercept form of the line as $y = -1x + 3$. Add x to both sides to obtain $x + y = 3$, the equation in standard form.

55) A Solve by adding -7 to each side of the inequality. Since the absolute value of x is less than 6, x must be between -6 and 6. The end points are not included so the circles on the graph are hollow.

56) D Be sure to enclose the sum of the number and 6 in parentheses.

57) B Let x = the speed of the boat in still water and c = the speed of the current.

	rate	time	distance
upstream	x - c	3	30
downstream	x + c	1.5	30

Solve the system:

$$3x - 3c = 30$$
$$1.5x + 1.5c = 30$$

58) B Each number in the domain can only be matched with one number in the range. A is not a function because 0 is mapped to 4 different numbers in the range. In C, 1 is mapped to two different numbers. In D, 4 is also mapped to two different numbers.

59) D Solve the denominator for 0. These values will be excluded from the domain.

$$2x^2 - 3 = 0$$
$$2x^2 = 3$$
$$x^2 = 3/2$$
$$x = \sqrt{\tfrac{3}{2}} = \sqrt{\tfrac{3}{2}} \bullet \sqrt{\tfrac{2}{2}} = \tfrac{\pm\sqrt{6}}{2}$$

60) C Glancing first at the solution choices, factor (y - x) from each term. This leaves -8 from the first term and a from the second term: $(a - 8)(y - x)$

61) D The complete factorization for a difference of cubes is $(k - m)(k^2 + mk + m2)$.

62) A Distribute and combine like terms to obtain $7x^2 - 14 = 0$. Add 14 to both sides, then divide by 7. Since $x^2 = 2$, $x = \sqrt{2}$

63) C Simplify each radical by factoring out the perfect squares:

$$5\sqrt{3} + 7\sqrt{3} - 4\sqrt{3} = 8\sqrt{3}$$

64) D The discriminate is the number under the radical sign. Since it is negative the two roots of the equation are complex.

65) B Since the vertex of the parabola is three units to the left, we choose the solution where 3 is subtracted from x, then the quantity is squared.

66) D The constant of variation for an inverse proportion is xy.

67) B y/x-216=x/18, Solve 36=6x.

68) C

69) A

70) C

71) C The angles in A are exterior. In B, the angles are vertical. The angles in D are consecutive, not adjacent.

72) A Each interior angle of the hexagon measures 120°. The isosceles triangle on the left has angles which measure 120, 30, and 30. By alternate interior angle theorem, $\angle 1$ is also 30.

73) C In any triangle, an exterior angle is equal to the sum of the remote interior angles.

74) B Use SAS with the last side being the vertical line common to both triangles.

75) D Angles formed by intersecting lines are called vertical angles and are congruent.

76) B In similar polygons, the areas are proportional to the squares of the sides. $6^2:8^2$; 36:64

77) A The sides are in the same ratio.

78) B The altitude from the right angle to the hypotenuse of any right triangle is the geometric mean of the two segments which are formed. Multiply 7 x 14 and take the square root.

79) A In a 30-60- 90 right triangle, the leg opposite the 30° angle is half the length of the hypotenuse.

80) D Minor arc AC measures 50°, the same as the central angle. To determine the measure of the major arc, subtract from 360.

81) C An inscribed angle is equal to one half the measure of the intercepted arc.

82) C The points marked C and D are the intersection of the circles with centers A and B.

83) C Using a compass, point K is found to be equidistant from A and B.

84) D A postulate is an accepted property of real numbers or geometric figures which cannot be proven, A, B. and C are theorems which can be proven.

85) B The point, line, and plane are the three undefined concepts on which plane geometry is based.

86) B To obtain the final side, add CD to both BC and ED.

87) D The isosceles triangle theorem states that the base angles are congruent, and the reflexive property states that every segment is congruent to itself.

88) C Using the distance formula

$$\sqrt{[3-(-3)]^2 + (7-4)^2}$$

$$= \sqrt{36+9}$$

$$= 3\sqrt{5}$$

89) D Using the midpoint formula

$x = (2 + 7)/2 \qquad y = (5 + -4)/2$

90) B

91) B

92) C

93) B

94) D

95) B

96) A

97) C

98) C

99) A

100) D

101) Let N = .636363.... Then multiplying both sides of the equation by 100 or 10^2 (because there are 2 repeated numbers), we get 100N = 63.636363... Then subtracting the two equations gives 99N = 63 or N = $\frac{63}{99}=\frac{7}{11}$.
Answer is B

102) I is not closed because $\frac{4}{.5}=8$ and 8 is not in the set.
III is not closed because $\frac{1}{0}$ is undefined.
II is closed because $\frac{-1}{1}=-1, \frac{1}{-1}=-1, \frac{1}{1}=1, \frac{-1}{-1}=1$ and all the answers are in the set. **Answer is B**

103) **Answer is D** because a + (-a) = 0 is a statement of the Additive Inverse Property of Algebra.

104) To find the inverse, $f^{-1}(x)$, of the given function, reverse the variables in the given equation, y = 3x – 2, to get x = 3y – 2. Then solve for y as follows: x+2 = 3y, and y = $\frac{x+2}{3}$. **Answer is D.**

105) Before the tax, the total comes to \$365.94. Then .065(365.94) = 23.79. With the tax added on, the total bill is 365.94 + 23.79 = \$389.73. (Quicker way: 1.065(365.94) = 389.73.) **Answer is A**

106) Recall: 30 days in April and 31 in March. 8 days in March + 30 days in April + 22 days in May brings him to a total of 60 days on May 22. **Answer is B.**

107) A composite number is a number which is not prime. The prime number sequence begins 2,3,5,7,11,13,17,.... To determine which of the expressions is always composite, experiment with different values of x and y, such as x=3 and y=2, or x=5 and y=2. It turns out that 5xy will always be an even number, and therefore, composite, if y=2. **Answer is C.**

108) Using FOIL to do the expansion, we get $(x + y^2)^2 = (x + y^2)(x + y^2) = x^2 + 2xy^2 + y^4$. **Answer is D**.

109) In scientific notation, the decimal point belongs to the right of the 4, the first significant digit. To get from 4.56 x 10^{-5} back to 0.0000456, we would move the decimal point 5 places to the left. **Answer is D.**

110) Area of triangle AOB is .5(5)(5) = 12.5 square meters. Since $\frac{90}{360} = .25$, the area of sector AOB (pie-shaped piece) is approximately .25(π)5^2 = 19.63. Subtracting the triangle area from the sector area to get the area of segment AB, we get approximately 19.63-12.5 = 7.13 square meters. **Answer is B.**

111) The formula for the volume of a cone is V = $\frac{1}{3}Bh$, (font)where B is the area of the circular base and h is the height. If the area of the base is tripled, the volume becomes V = $\frac{1}{3}(3B)h = Bh$(font), or three times the original area. **Answer is C**.

112) Divide the figure into 2 rectangles and one quarter circle. The tall rectangle on the left will have dimensions 10 by 4 and area 40. The rectangle in the center will have dimensions 7 by 10 and area 70. The quarter circle will have area .25(π)7^2 = 38.48. The total area is therefore approximately 148.48. **Answer is B.**

113) Since an ordinary cookie would not weigh as much as 1 kilogram, or as little as 1 gram or 15 milligrams, the only reasonable answer is 15 grams. **Answer is C.**

114) Arrange the data in ascending order: 12,13,14,16,17,19. The median is the middle value in a list with an odd number of entries. When there are an even number of entries, the median is the mean of the two center entries. Here the average of 14 and 16 is 15. **Answer is C.**

115) In this set of data, the median (see #14) would be the most representative measure of central tendency, since the median is independent of extreme values. Because of the 10% outlier, the mean (average) would be disproportionately skewed. In this data set, it is true that the median and the mode (number which occurs most often) are the same, but the median remains the best choice because of its special properties. **Answer is B.**

116) In kindergarten, first grade, and third grade, there are more boys than girls. The number of extra girls in grade two is more than made up for by the extra boys in all the other grades put together. **Answer is B.**

117) The values of 5 and –5 must be omitted from the domain of all real numbers because if x took on either of those values, the denominator of the fraction would have a value of 0, and therefore the fraction would be undefined. **Answer is B.**

118) By observation, we see that the graph has a y-intercept of 2 and a slope of 2/1 = 2. Therefore its equation is y = mx + b = 2x + 2. Rearranging the terms gives 2x – y = -2. **Answer is B.**

119) Using the Distributive Property and other properties of equality to isolate v_0 gives $d = atv_t - atv_0$, $atv_0 = atv_t - d$, $v_0 = \frac{atv_t - d}{at}$. **Answer is D.**

120) Removing the common factor of 6 and then factoring the sum of two cubes gives $6 + 48m^3 = 6(1 + 8m^3) = 6(1 + 2m)(1^2 - 2m + (2m)^2)$. **Answer is A.**

121) B is not the graph of a function. D is the graph of a parabola where the coefficient of x^2 is negative. A appears to be the graph of $y = x^2$. To find the x-intercepts of $y = x^2 + 3x$, set y = 0 and solve for x: $0 = x^2 + 3x = x(x + 3)$ to get x = 0 or x = -3. Therefore, the graph of the function intersects the x-axis at x=0 and x=-3. **Answer is C.**

122) Set up the direct variation: $\frac{V}{r^2} = \frac{V}{r^2}$. Substituting gives $\frac{80}{16} = \frac{V}{9}$. Solving for V gives 45 liters per minute. **Answer is A.**

123) Multiplying equation 1 by 2, and equation 2 by –3, and then adding together the two resulting equations gives -11y + 22z = 0. Solving for y gives y = 2z. In the meantime, multiplying equation 3 by –2 and adding it to equation 2 gives –y – 12z = -14. Then substituting 2z for y, yields the result z = 1. Subsequently, one can easily find that y = 2, and x = -1. **Answer is A.**

124) Using the definition of absolute value, two equations are possible: 18 = 4 + 2x or 18 = 4 – 2x. Solving for x gives x = 7 or x = -7. **Answer is C.**

125) Rewriting the inequality gives $x^2 - 5x + 6 > 0$. Factoring gives $(x - 2)(x - 3) > 0$. The two cut-off points on the number line are now at x = 2 and x = 3. Choosing a random number in each of the three parts of the number line, we test them to see if they produce a true statement. If x = 0 or x = 4, (x-2)(x-3)>0 is true. If x = 2.5, (x-2)(x-3)>0 is false. Therefore the solution set is all numbers smaller than 2 or greater than 3. **Answer is D.**

126) Possible rational roots of the equation $0 = x^3 + x^2 - 14x - 24$ are all the positive and negative factors of 24. By substituting into the equation, we find that –2 is a root, and therefore that x+2 is a factor. By performing the long division $(x^3 + x^2 - 14x - 24)/(x+2)$, we can find that another factor of the original equation is $x^2 - x - 12$ or (x-4)(x+3). Therefore the zeros of the original function are –2, -3, and 4. **Answer is D.**

127) Getting the bases the same gives us $3^{\frac{1}{2}}3^{\frac{2}{3}}$. Adding exponents gives $3^{\frac{7}{6}}$. Then some additional manipulation of exponents produces $3^{\frac{7}{6}} = 3^{\frac{14}{12}} = (3^2)^{\frac{7}{12}} = 9^{\frac{7}{12}}$. **Answer is B.**

128) Simplifying radicals gives $\sqrt{27}+\sqrt{75}=3\sqrt{3}+5\sqrt{3}=8\sqrt{3}$. **Answer is A.**

129) Multiplying numerator and denominator by the conjugate gives
$\frac{10}{1+3i}\times\frac{1-3i}{1-3i}=\frac{10(1-3i)}{1-9i^2}=\frac{10(1-3i)}{1-9(-1)}=\frac{10(1-3i)}{10}=1-3i$. **Answer is D.**

130) To find the 100th term: t_{100} = -6 + 99(4) = 390. To find the sum of the first 100 terms: S = $\frac{100}{2}(-6+390)=19200$. **Answer is A.**

131) There are 3 slots to fill. There are 3 choices for the first, 7 for the second, and 6 for the third. Therefore, the total number of choices is 3(7)(6) = 126. **Answer is A.**

132) The set-up for finding the seventh term is $\frac{8(7)(6)(5)(4)(3)}{6(5)(4)(3)(2)(1)}(2a)^{8-6}b^6$ which gives $28(4a^2b^6)$ or $112a^2b^6$. **Answer is C.**

133) By definition, parallel lines are coplanar lines without any common points. **Answer is B.**

134) A set of n objects has 2^n subsets. Therefore, here we have 2^4 = 16 subsets. These subsets include four which have only 1 element each, six which have 2 elements each, four which have 3 elements each, plus the original set, and the empty set. **Answer is B.**

135) Formula for finding the measure of each interior angle of a regular polygon with n sides is $\frac{(n-2)180}{n}$. For n=10, we get $\frac{8(180)}{10}=144$. **Answer is C.**

136) Draw a right triangle with legs of 6 and 8. Find the hypotenuse using the Pythagorean Theorem. $6^2 + 8^2 = c^2$. Therefore, c = 10 miles. **Answer is B.**

137) The formula relating the measure of angle K and the two arcs it intercepts is $m\angle K=\frac{1}{2}(mPS-mAD)$. Substituting the known values, we get $10=\frac{1}{2}(40-mAD)$. Solving for mAD gives an answer of 20 degrees. **Answer is B.**

138) Given a point on a line, place the compass point there and draw two arcs intersecting the line in two points, one on either side of the given point. Then using any radius larger than half the new segment produced, and with the pointer at each end of the new segment, draw arcs which intersect above the line. Connect this new point with the given point. **Answer is D.**

139) By definition this describes the procedure of an indirect proof. **Answer is C.**

140) Since side AK is common to both triangles, the triangles can be proved congruent by using the Side-Angle-Side Postulate. **Answer is C.**

141) In an ordinary parallelogram, the diagonals are not perpendicular or equal in length. In a rectangle, the diagonals are not necessarily perpendicular. In a rhombus, the diagonals are not equal in length. In a square, the diagonals are both perpendicular and congruent. **Answer is C.**

142) The most one can say with certainty is that the median (segment drawn to the midpoint of the opposite side) and the altitude (segment drawn perpendicular to the opposite side) of a triangle may coincide, but they more often do not. In an isosceles triangle, the median and the altitude to the base are the same segment. **Answer is A.**

143) In the mid-nineteenth century, Reimann and other mathematicians developed elliptic geometry. **Answer is B.**

144) Let's assume the base of the rectangular solid (box) is 3 by 4, and the height is 5. Then the surface area of the top and bottom together is 2(12) = 24. The sum of the areas of the front and back are 2(15) = 30, while the sum of the areas of the sides are 2(20)=40. The total surface area is therefore 94 square feet. **Answer is C.**

145) Find the area of the garden and then subtract the area of the fountain: $30(60)-\pi(5)^2$ or approximately 1721 square meters. **Answer is A.**

146) To find the area of the shaded region, find the area of triangle ABC and then subtract the area of triangle DBE. The area of triangle ABC is .5(6x)(y) = 3xy. The area of triangle DBE is .5(2x)(y) = xy. The difference is 2xy. **Answer is B.**

SUBAREA VI. HISTORY, PHILOSOPHY, AND METHODOLOGY OF SCIENCE

Competency 0021 Understand the nature of scientific thought and inquiry and the historical development of major scientific ideas

The combination of science, mathematics and technology forms the scientific endeavor and makes science a success. It is impossible to study science on its own without the support of other disciplines like mathematics, technology, geology, physics, and other disciplines as well.

Science is tentative. By definition, it is searching for information by making educated guesses. It must be replicable. Another scientist must be able to achieve the same results under the same conditions at a later time. The term empirical means it must be assessed through tests and observations. Science changes over time. Science is limited by the available technology. An example of this would be the relationship of the discovery of the cell and the invention of the microscope. As our technology improves, more hypotheses will become theories and possibly laws. Science is also limited by the data that is able to be collected. Data may be interpreted differently on different occasions. Science limitations cause explanations to be changeable as new technologies emerge. New technologies gather previously unavailable data and enable us to build upon current theories with new information.

Ancient history followed the geocentric theory, which was displaced by the heliocentric theory developed by Copernicus, Ptolemy and Kepler. Newton's laws of motion by Sir Isaac Newton were based on mass, force and acceleration, and state that the force of gravity between any two objects in the universe depends upon their mass and distance. These laws are still widely used today. In the 20th century, Albert Einstein was the most outstanding scientist for his work on relativity, which led to his theory that $E=mc^2$. Early in the 20th century, Alfred Wegener proposed his theory of continental drift, stating that continents moved away from the super continent, Pangaea. This theory was accepted in the 1960s when more evidence was collected to this. John Dalton and Lavosier made significant contributions in the fields of atom and matter. The Curies and Ernest Rutherford contributed greatly to radioactivity and the splitting of atom, which have lot of practical applications. Charles Darwin proposed his theory of evolution and Gregor Mendel's experiments on peas helped us to understand heredity. The most significant improvement was the industrial revolution in Britain, in which science was applied practically to increase the productivity and also introduced a number of social problems like child labor.

The nature of science mainly consists of three important things:

1. The scientific world view

This includes some very important issues like – it is possible to understand this highly organized world and its complexities with the help of latest technology. Scientific ideas are subject to change. After repeated experiments, a theory is established, but this theory could be changed or supported in the future. Only laws that occur naturally do not change. Scientific knowledge may not be discarded but is modified – e.g., Albert Einstein didn't discard the Newtonian principles but modified them in his theory of relativity. Also, science can't answer all of our questions. We can't find answers to questions related to our beliefs, moral values, and our norms.

2. Scientific inquiry

Scientific inquiry starts with a simple question. This simple question leads to information gathering, an educated guess otherwise known as a hypothesis. To prove the hypothesis, an experiment has to be conducted, which yields data and the conclusion. All experiments must be repeated at least twice to get reliable results. Thus scientific inquiry leads to new knowledge or verifying established theories. Science requires proof or evidence. Science is dependent on accuracy, not bias or prejudice. In science, there is no place for preconceived ideas or premeditated results. By using their senses and modern technology, scientists will be able to get reliable information. Science is a combination of logic and imagination. A scientist needs to think and imagine and be able to reason. Science explains, reasons and predicts. These three are interwoven and are inseparable. While reasoning is absolutely important for science, there should be no bias or prejudice. Science is not authoritarian because it has been shown that scientific authority can be wrong. No one can determine or make decisions for others on any issue.

3. Scientific enterprise

Science is a complex activity involving various people and places. A scientist may work alone or in a laboratory, classroom or for that matter, anywhere. Mostly it is a group activity requiring lot of social skills of cooperation, communication of results or findings, consultations, discussions etc. Science demands a high degree of communication to the governments, funding authorities and to public.

Scientific hypothesis are subject to experimental and observational confirmation

Science is a process of checks and balances. It is expected that scientific findings will be challenged, and in many cases, retested. Often one experiment will be the beginning point for another. While bias does exist, the use of controlled experiments and an awareness on the part of the scientist can go far in ensuring a sound experiment. Even if the science is well done, it may still be questioned. It is through this continual search that hypotheses are made into theories, and sometimes become laws. It is also through this search that new information is discovered.

Historical overview

The history of biology follows man's understanding of the living world from the earliest recorded history to modern times. Though the concept of biology as a field of science arose only in the 19th century, its origins could be traced back to the ancient Greeks (Galen and Aristotle).

During the Renaissance and Age of Discovery, renewed interest in the rapidly increasing number of known organisms generated lot of interest in biology.

Andreas Vesalius (1514-1564) was a Belgian anatomist and physician whose dissections of the human body and the descriptions of his findings helped to correct the misconceptions of science. The books Vesalius wrote on anatomy were the most accurate and comprehensive anatomical texts of time.

Anton van Leeuwenhoek is known as the father of microscopy. In the 1650s, Leeuwenhoek began making tiny lenses that gave magnifications up to 300x. He was the first to see and describe bacteria, yeast plants, and the microscopic life found in water. Over the years, light microscopes have advanced to produce greater clarity and magnification. The transmission electron microscope (TEM) was developed in the 1950s. Instead of light, a beam of electrons passes through the specimen. Transmission electron microscopes have a resolution about one thousand times greater than light microscopes. The disadvantage of the TEM is that the chemical and physical methods used to prepare the sample result in the death of the specimen.

Carl Von Linnaeus (1707-1778), a Swedish botanist, physician, and zoologist, is well known for his contributions in ecology and taxonomy. Linnaeus is famous for his binomial system of nomenclature in which each living organism has two names, a genus and a species name. He is considered the father of modern ecology and taxonomy.

In the late 1800s, Pasteur discovered the role of microorganisms in the cause of disease, pasteurization, and the rabies vaccine. Koch took his observations one step further by postulating that specific diseases were caused by specific pathogens. **Koch's postulates** are still used as guidelines in the field of microbiology. They state that the same pathogen must be found in every diseased person, the pathogen must be isolated and grown in culture, the disease must be induced in experimental animals from the culture, and the same pathogen must be isolated from the experimental animal.

In the 18th century, many fields of science like botany, zoology and geology began to evolve as scientific disciplines in the modern sense.

In the 20th century, the rediscovery of Mendel's work led to the rapid development of genetics by Thomas Hunt Morgan and his students.

DNA structure was another key event in biological study. In the 1950s, James Watson and Francis Crick discovered the structure of a DNA molecule as that of a double helix. This structure made it possible to explain DNA's ability to replicate and to control the synthesis of proteins.

Following the cracking of the genetic code, biology has largely split between organismal biology-consisting of ecology, ethology, systematics, paleontology, evolutionary biology, developmental biology, and other disciplines that deal with whole organisms or group of organisms, and the disciplines related to molecular biology, which include cell biology, biophysics, biochemistry, neuroscience, and immunology.

The use of animals in biological research has expedited many scientific discoveries. Animal research has allowed scientists to learn more about animal biological systems, including the circulatory and reproductive systems. One significant use of animals is for the testing of drugs, vaccines, and other products (such as perfumes and shampoos) before use or consumption by humans. There are both significant pros and cons of animal research. The debate about the ethical treatment of animals has been ongoing since the introduction of animals to research. Many people believe the use of animals in research is cruel and unnecessary. Animal use is federally and locally regulated. The purpose of the Institutional Animal Care and Use Committee (IACUC) is to oversee and evaluate all aspects of an institution's animal care and use program.

Competency 0022 Understand the principles and procedures of research and experimental design

The scientific method is the basic process behind science. It involves several steps beginning with hypothesis formulation and working through to the conclusion.

Posing a question
Although many discoveries happen by chance, the standard thought process of a scientist begins with forming a question to research. The more limited the question, the easier it is to set up an experiment to answer it.

Form a hypothesis
Once the question is formulated, take an educated guess about the answer to the problem or question. This 'best guess' is your hypothesis.

Conducting the test
To make a test fair, data from an experiment must have a **variable** or any condition that can be changed such as temperature or mass. A good test will try to manipulate as few variables as possible so as to see which variable is responsible for the result. This requires a second example of a **control**. A control is an extra setup in which all the conditions are the same and the variable being tested is unchanged. A comparison of the control and the test is done to determine the result of changing the variable in the test.

Observe and record the data
Reporting of the data should state specifics of how the measurements were calculated. A graduated cylinder needs to be read with proper procedures. As beginning students, technique must be part of the instructional process so as to give validity to the data.

Drawing a conclusion
After recording data, you compare your data with that of other groups. A conclusion is the judgment derived from the data results.

Normally, knowledge is integrated in the form of a lab report. A report has many sections. It should include a specific **title** and tell exactly what is being studied. The **abstract** is a summary of the report written at the beginning of the paper. The **purpose** should always be defined and will state the problem. The purpose should include the **hypothesis** (educated guess) of what is expected from the outcome of the experiment. The entire experiment should relate to this problem. It is important to describe exactly what was done to prove or disprove a hypothesis. A **control** is necessary to prove that the results occurred from the changed conditions and would not have happened normally. Only one variable should be manipulated at a time. **Observations** and **results** of the experiment should be recorded including all results from data. Drawings, graphs and illustrations should be included to support information. Observations are objective, whereas analysis and interpretation is subjective. A **conclusion** should explain why the results of the experiment either proved or disproved the hypothesis.

A scientific theory is an explanation of a set of related observations based on a proven hypothesis. A scientific law is an explanation of events that occur with uniformity under the same conditions (laws of nature, law of gravitation).

Competency 0023 Understand the procedures for gathering, organizing, interpreting, evaluating, and communicating scientific information

Use appropriate methods, tools, and technologies for gathering, recording, processing, analyzing, and evaluating data and for communicating the results of scientific investigations

The procedure used to obtain data is important to the outcome. Experiments consist of **controls** and **variables**. A control is the experiment run under normal conditions. The variable includes a factor that is changed. In biology, the variable may be light, temperature, pH, time, etc. The differences in tested variables may be used to make a prediction or form a hypothesis. Only one variable should be tested at a time. One would not alter both the temperature and pH of the experimental subject.

An **independent variable** is one that is changed or manipulated by the researcher. This could be the amount of light given to a plant or the temperature at which bacteria is grown. The **dependent variable** is that which is influenced by the independent variable.

Measurements may be taken in different ways. There is an appropriate measuring device for each aspect of biology. A graduated cylinder is used to measure volume. A balance is used to measure mass. A microscope is used to view microscopic objects. A centrifuge is used to separate two or more parts in a liquid sample. The list goes on, but you get the point. For each variable, there is an appropriate way to measure it. The internet and teaching guides are virtually unlimited resources for laboratory ideas. You should be imparting on the students the importance of the method with which they conduct the study, the resource they use to do so, the concept of double checking their work, and the use of appropriate units.

Biologists use a variety of tools and technologies to perform tests, collect and display data, and analyze relationships. Examples of commonly used tools include computer-linked probes, spreadsheets, and graphing calculators.

Biologists use computer-linked probes to measure various environmental factors including temperature, dissolved oxygen, pH, ionic concentration, and pressure. The advantage of computer-linked probes, as compared to more traditional observational tools, is that the probes automatically gather data and present it in an accessible format. This property of computer-linked probes eliminates the need for constant human observation and manipulation.

Biologists use spreadsheets to organize, analyze, and display data. For example, conservation ecologists use spreadsheets to model population growth and development, apply sampling techniques, and create statistical distributions to analyze relationships. Spreadsheet use simplifies data collection and manipulation and allows the presentation of data in a logical and understandable format.

Graphing calculators are another technology with many applications to biology. For example, biologists use algebraic functions to analyze growth, development and other natural processes. Graphing calculators can manipulate algebraic data and create graphs for analysis and observation. In addition, biologists use the matrix function of graphing calculators to model problems in genetics. The use of graphing calculators simplifies the creation of graphical displays including histograms, scatter plots, and line graphs. Biologists can also transfer data and displays to computers for further analysis. Finally, biologists connect computer-linked probes, used to collect data, to graphing calculators to ease the collection, transmission, and analysis of data.

Select appropriate methods and criteria for organizing and displaying data (e.g., tables, graphs, models)

The type of graphic representation used to display observations depends on the data that is collected. **Line graphs** are used to compare different sets of related data or to predict data that has not yet been measured. An example of a line graph would be comparing the rate of activity of different enzymes at varying temperatures. A **bar graph** or **histogram** is used to compare different items and make comparisons based on this data. An example of a bar graph would be comparing the ages of children in a classroom. A **pie chart** is useful when organizing data as part of a whole. A good use for a pie chart would be displaying the percent of time students spend on various after school activities.

As noted before, the independent variable is controlled by the experimenter. This variable is placed on the x-axis (horizontal axis). The dependent variable is influenced by the independent variable and is placed on the y-axis (vertical axis). It is important to choose the appropriate units for labeling the axes. It is best to take the largest value to be plotted and divide it by the number of blocks, and rounding to the nearest whole number.

Demonstrate an understanding of concepts of precision, accuracy, and error with regard to gathering and recording scientific data

Accuracy is the degree of conformity of a measured, calculated quantity to its actual (true) value. Precision also called reproducibility or repeatability and is the degree to which further measurements or calculations will show the same or similar results.

Accuracy is the degree of veracity while precision is the degree of reproducibility.

The best analogy to explain accuracy and precision is the target comparison.

Repeated measurements are compared to arrows that are fired at a target. Accuracy describes the closeness of arrows to the bull's eye at the target center. Arrows that strike closer to the bull's eye are considered more accurate. While precision is how many arrows cluster together on the target irregardless whether they strike the target.

All experimental uncertainty is due to either random errors or systematic errors.

Random errors are statistical fluctuations in the measured data due to the precision limitations of the measurement device. Random errors usually result from the experimenter's inability to take the same measurement in exactly the same way to get exactly the same number.

Systematic errors, by contrast, are reproducible inaccuracies that are consistently in the same direction. Systematic errors are often due to a problem that persists throughout the entire experiment.

Systematic and random errors refer to problems associated with making measurements. Mistakes made in the calculations or in reading the instrument are not considered in error analysis.

Demonstrate knowledge of the measurement units used in scientific investigations

Science uses the **metric system**; as it is accepted worldwide and allows easier comparison among experiments done by scientists around the world.

The meter is the basic metric unit of length. One meter is 1.1 yards. The liter is the basic metric unit of volume. 1 gallon is 3.846 liters. The gram is the basic metric unit of mass. 1000 grams is 2.2 pounds.

deca- 10X the base unit
hecto- 100X the base unit
kilo- 1,000X the base unit
mega- 1,000,000X the base unit
giga- 1,000,000,000X the base unit
tera- 1,000,000,000,000X the base unit

deci - 1/10 the base unit
centi - 1/100 the base unit
milli - 1/1,000 the base unit
micro- 1/1,000,000 the base unit
nano- 1/1,000,000,000 the base unit
pico- 1/1,000,000,000,000 the base

The common instrument used for measuring volume is the graduated cylinder. The unit of measurement is usually in milliliters (mL). It is important for accurate measure to read the liquid in the cylinder at the bottom of the meniscus, the curved surface of the liquid.

The common instrument used is measuring mass is the triple beam balance. The triple beam balance is measured in as low as tenths of a gram and can be estimated to the hundredths of a gram.

The ruler or meter sticks are the most commonly used instruments for measuring length. Measurements in science should always be measured in metric units. Be sure when measuring length that the metric units are used.

Identify and evaluate various sources of scientific information

Because people often attempt to use scientific evidence in support of political or personal agendas, the ability to evaluate the credibility of scientific claims is a necessary skill in today's society. In evaluating scientific claims made in the media, public debates, and advertising, one should follow several guidelines.

First, scientific, peer-reviewed journals are the most accepted source for information on scientific experiments and studies. One should carefully scrutinize any claim that does not reference peer-reviewed literature.

Second, the media and those with an agenda to advance (advertisers, debaters, etc.) often overemphasize the certainty and importance of experimental results. One should question any scientific claim that sounds fantastical or overly certain.

Finally, knowledge of experimental design and the scientific method is important in evaluating the credibility of studies. For example, one should look for the inclusion of control groups and the presence of data to support the given conclusions.

Competency 0024 Understand the safe and proper use of tools, equipment, and materials (including chemicals and living organisms) related to classroom and other science investigations

Bunsen burners - Hot plates should be used whenever possible to avoid the risk of burns or fire. If Bunsen burners are used, the following precautions should be followed:

1. Know the location of fire extinguishers and safety blankets and train students in their use. Long hair and long sleeves should be secured and out of the way.
2. Turn the gas all the way on and make a spark with the striker. The preferred method to light burners is to use strikers rather than matches.
3. Adjust the air valve at the bottom of the Bunsen burner until the flame shows an inner cone.
4. Adjust the flow of gas to the desired flame height by using the adjustment valve.
5. Do not touch the barrel of the burner (it is hot).

Graduated Cylinder - These are used for precise measurements. They should always be placed on a flat surface. The surface of the liquid will form a meniscus (lens-shaped curve). The measurement is read at the bottom of this curve.

Balance - Electronic balances are easier to use, but more expensive. An electronic balance should always be tared (returned to zero) before measuring and used on a flat surface. Substances should always be placed on a piece of paper to avoid spills and/or damage to the instrument. Triple beam balances must be used on a level surface. There are screws located at the bottom of the balance to make any adjustments. Start with the largest counterweight first and proceed toward the last notch that does not tip the balance. Do the same with the next largest, etc until the pointer remains at zero. The total mass is the total of all the readings on the beams. Again, use paper under the substance to protect the equipment.

Buret – A buret is used to dispense precisely measured volumes of liquid. A stopcock is used to control the volume of liquid being dispensed at a time.

Light microscopes are commonly used in laboratory experiments. Several procedures should be followed to properly care for this equipment:

- Clean all lenses with lens paper only.
- Carry microscopes with two hands; one on the arm and one on the base.
- Always begin focusing on low power, then switch to high power.
- Store microscopes with the low power objective down.
- Always use a coverslip when viewing wet mount slides.
- Bring the objective down to its lowest position then focus by moving up to avoid breaking the slide or scratching the lens.

Wet mount slides should be made by placing a drop of water on the specimen and then putting a glass coverslip on top of the drop of water. Dropping the coverslip at a forty-five degree angle will help in avoiding air bubbles. Total magnification is determined by multiplying the ocular (usually 10X) and the objective (usually 10X on low, 40X on high).

All laboratory solutions should be prepared as directed in the lab manual. Care should be taken to avoid contamination. All glassware should be rinsed thoroughly with distilled water before using, and cleaned well after use. Safety goggles should be worn while working with glassware in case of an accident. All solutions should be made with distilled water as tap water contains dissolved particles that may affect the results of an experiment. Chemical storage should be located in a secured, dry area. Chemicals should be stored in accordance with reactability. Acids are to be locked in a separate area. Used solutions should be disposed of according to local disposal procedures. Any questions regarding safe disposal or chemical safety may be directed to the local fire department.

Chromatography uses the principles of capillary action to separate substances such as plant pigments. Molecules of a larger size will move slower up the paper, whereas smaller molecules will move more quickly producing lines of pigments.

Spectrophotometry uses percent light absorbance to measure a color change, thus giving qualitative data a quantitative value.

Centrifugation involves spinning substances at a high speed. The more dense part of a solution will settle to the bottom of the test tube, while the lighter material will stay on top. Centrifugation is used to separate blood into blood cells and plasma, with the heavier blood cells settling to the bottom.

Electrophoresis uses electrical charges of molecules to separate them according to their size. The molecules, such as DNA or proteins, are pulled through a gel towards either the positive end of the gel box (if the material has a negative charge) or the negative end of the gel box (if the material has a positive charge).

Computer technology has greatly improved the collection and interpretation of scientific data. Molecular findings have been enhanced through the use of computer images. Technology has revolutionized access to data via the internet and shared databases. The manipulation of data is enhanced by sophisticated software capabilities. Computer engineering advances have produced such products as MRIs and CT scans in medicine. Laser technology has numerous applications with refining precision.

Satellites have improved our ability to communicate and transmit radio and television signals. Navigational abilities have been greatly improved through the use of satellite signals. Sonar uses sound waves to locate objects and is especially useful underwater. The sound waves bounce off the object and are used to assist in location. Seismographs record vibrations in the earth and allow us to measure earthquake activity.

Evaluate equipment, materials, procedures, and settings for potential safety hazards

Safety in the science classroom and laboratory is of paramount importance to the science educator. The following is a general summary of the types of safety equipment that should be made available within a given school system as well as general locations where the protective equipment or devices should be maintained and used. Please note that this is only a partial list and that your school system should be reviewed for unique hazards and site-specific hazards at each facility.

The key to maintaining a safe learning environment is through proactive training and regular in-service updates for all staff and students who utilize the science laboratory. Proactive training should include how to **identify potential hazards**, **evaluate potential hazards**, and **how to prevent or respond to hazards**. The following types of training should be considered:

a) Right to Know (OSHA training on the importance and benefits of properly recognizing and safely working with hazardous materials) along with some basic chemical hygiene as well as how to read and understand a material safety data sheet,
b) instruction in how to use a fire extinguisher,
c) instruction in how to use a chemical fume hood,
d) general guidance in when and how to use personal protective equipment (e.g. safety glasses or gloves), and
e) instruction in how to monitor activities for potential impacts on indoor air quality.

It is also important for the instructor to utilize **Material Data Safety Sheets**. Maintain a copy of the material safety data sheet for every item in your chemical inventory. This information will assist you in determining how to store and handle your materials by outlining the health and safety hazards posed by the substance. In most cases the manufacturer will provide recommendations with regard to protective equipment, ventilation and storage practices. This information should be your first guide when considering the use of a new material.

Frequent monitoring and in-service training on all equipment, materials, and procedures will help to ensure a safe and orderly laboratory environment. It will also provide everyone who uses the laboratory the safety fundamentals necessary to discern a safety hazard and to respond appropriately.

Safe practices and procedures in all areas related to science instruction

In addition to requirements set forth by your place of employment, the NABT (National Association of Biology Teachers) and ISEF (International Science Education Foundation) have been instrumental in setting parameters for the science classroom. All science labs should contain the following items of **safety equipment**. Those marked with an asterisk are requirements by state laws.

    * fire blanket which is visible and accessible
    *Ground Fault Circuit Interrupters (GCFI) within two feet of water supplies
    *signs designating room exits
    *emergency shower providing a continuous flow of water
    *emergency eye wash station which can be activated by the foot or forearm
    *eye protection for every student and a means of sanitizing equipment
    *emergency exhaust fans providing ventilation to the outside of the building
    *master cut-off switches for gas, electric and compressed air. Switches must have permanently attached handles. Cut-off switches must be clearly labeled.
    *an ABC fire extinguisher
    *storage cabinets for flammable materials
    -chemical spill control kit
    -fume hood with a motor which is spark proof
    -protective laboratory aprons made of flame retardant material
    -signs which will alert potential hazardous conditions
    -containers for broken glassware, flammables, corrosives, and waste
    -containers should be labeled.

Students should wear safety goggles when performing dissections, heating, or while using acids and bases. Hair should always be tied back and objects should never be placed in the mouth. Food should not be consumed while in the laboratory. Hands should always be washed before and after laboratory experiments. In case of an accident, eye washes and showers should be used for eye contamination or a chemical spill that covers the student's body. Small chemical spills should only be contained and cleaned by the teacher. Kitty litter or a chemical spill kit should be used to clean spill. For large spills, the school administration and the local fire department should be notified. Biological spills should also be handled only by the teacher. Contamination with biological waste can be cleaned by using bleach when appropriate.

Accidents and injuries should always be reported to the school administration and local health facilities. The severity of the accident or injury will determine the course of action to pursue.

It is the responsibility of the teacher to provide a safe environment for their students. Proper supervision greatly reduces the risk of injury and a teacher should never leave a class for any reason without providing alternate supervision. After an accident, two factors are considered; **foreseeability** and **negligence**. Foreseeability is the anticipation that an event may occur under certain circumstances. Negligence is the failure to exercise ordinary or reasonable care. Safety procedures should be a part of the science curriculum and a well-managed classroom is important to avoid potential lawsuits.

First response procedures, including first aid, for responding to accidents

All students and staff should be trained in first aid in the science classroom and laboratory. Please remember to always report all accidents, however minor, to the lab instructor immediately. In most situations 911 should immediately be called. Please refer to your school's specific safety plan for accidents in the classroom and laboratory. The classroom/laboratory should have a complete first-aid kit with supplies that are up-to-date and checked frequently for expiration.

Know the location and use of fire extinguishers, eye-wash stations, and safety showers in the lab.

Do not attempt to smother a fire in a beaker or flask with a fire extinguisher. The force of the stream of material from it will turn over the vessel and result in a bigger fire. Just place a watch glass or a wet towel over the container to cut off the supply of oxygen.

If your clothing is on fire, **do not run** because this only increases the burning. It is normally best to fall on the floor and roll over to smother the fire. If a student, whose clothing is on fire panics and begins to run, attempt to get the student on the floor and roll over to smother the flame. If necessary, use the fire blanket or safety shower in the lab to smother the fire.

Students with long hair should put their hair in a bun or a pony-tail to avoid their hair catching fire.

Below are common accidents that everyone who uses the laboratory should be trained in how to respond:

Burns (Chemical or Fire) – Use deluge shower for 15 minutes.

Burns (Clothing on fire) – Use safety shower immediately. Keep victim immersed 15 minutes to wash away both heat and chemicals. All burns should be examined by medical personnel.

Chemical spills – Chemical spills on hands or arms should be washed immediately with soap and water. Washing hands should become an instinctive response to any chemical spilled on hands. Spills that cover clothing and other parts of the body should be drenched under the safety shower. If strong acids or bases are spilled on clothing, the clothing should be removed. If a large area is affected, remove clothing and immerse victim in the safety shower. If a small area is affected, remove article of clothing and use a deluge shower for 15 minutes.

Eyes (chemical contamination) – Hold the eye wide open and flush with water from the eye wash for about 15 minutes. Seek medical attention.

Ingestion of chemicals or poisoning – See antidote chart on wall of lab for general first-aid directions. The victim should drink large amounts of water. All chemical poisonings should receive medical attention.

Information about safety, legal issues, and the proper use, storage, and proper disposal of scientific materials

Storing, identifying, and disposing of chemicals and biological materials

All laboratory solutions should be prepared as directed in the lab manual. Care should be taken to avoid contamination. All glassware should be rinsed thoroughly with distilled water before using and cleaned well after use. All solutions should be made with distilled water as tap water contains dissolved particles that may affect the results of an experiment. Unused solutions should be disposed of according to local disposal procedures.

The "Right to Know Law" covers science teachers who work with potentially hazardous chemicals. Briefly, the law states that employees must be informed of potentially toxic chemicals. An inventory must be made available if requested. The inventory must contain information about the hazards and properties of the chemicals. This inventory is to be checked against the "Substance List". Training must be provided on the safe handling and interpretation of the Material Safety Data Sheet.

The following chemicals are potential carcinogens and not allowed in school facilities: Acrylonitriel, Arsenic compounds, Asbestos, Bensidine, Benzene, Cadmium compounds, Chloroform, Chromium compounds, Ethylene oxide, Ortho-toluidine, Nickel powder, and Mercury.

Chemicals should not be stored on bench tops or heat sources. They should be stored in groups based on their reactivity with one another and in protective storage cabinets. All containers within the lab must be labeled. Suspected and known carcinogens must be labeled as such and segregated within trays to contain leaks and spills.

Chemical waste should be disposed of in properly labeled containers. Waste should be separated based on their reactivity with other chemicals. Biological material should never be stored near food or water used for human consumption. All biological material should be appropriately labeled. All blood and body fluids should be put in a well-contained container with a secure lid to prevent leaking. All biological waste should be disposed of in biological hazardous waste bags.

In addition to the safety laws set forth by the government for equipment necessary to the lab, OSHA (Occupational Safety and Health Administration) has helped to make environments safer by instituting signs that are bilingual. These signs use pictures rather than/in addition to words and feature eye-catching colors. Some of the best-known examples are exit, restrooms, and handicap accessible.

Of particular importance to laboratories are diamond safety signs, prohibitive signs, and triangle danger signs. Each sign encloses a descriptive picture.

As a teacher, you should utilize a MSDS (Material Safety Data Sheet) whenever you are preparing an experiment. It is designed to provide people with the proper procedures for handling or working with a particular substance. An MSDS includes information such as physical data (melting point, boiling point, etc.), toxicity, health effects, first aid, reactivity, storage, disposal, protective gear, and spill/leak procedures. These are particularly important if a spill or other accident occurs. You should review a few, available commonly online, and understand the listing procedures. Material safety data sheets are available directly from the company of acquisition or the internet. The manuals for equipment used in the lab should be read and understood before using them.

Knowledge of the safe and proper use of the scientific tools, equipment, chemicals, materials, and technology in scientific inquiry

Light microscopes are commonly used in high school laboratory experiments. Total magnification is determined by multiplying the ocular (usually 10X) and the objective (usually 10X on low, 40X on high) lenses. Several procedures should be followed to properly care for this equipment.

-Clean all lenses with lens paper only.
-Carry microscopes with two hands; one on the arm and one on the base.
-Always begin focusing on low power, then switch to high power.
-Store microscopes with the low power objective down.
-Always use a coverslip when viewing wet mount slides.
-Bring the objective down to its lowest position then focus moving up to avoid breaking the slide or scratching the lens.

Wet mount slides should be made by placing a drop of water on the specimen and then putting a glass coverslip on top of the drop of water. Dropping the coverslip at a forty-five degree angle will help in avoiding air bubbles.

Chromatography uses the principles of capillary action to separate substances such as plant pigments. Molecules of a larger size will move slower up the paper, whereas smaller molecules will move more quickly producing lines of pigment.

An **indicator** is any substance used to assist in the classification of another substance. An example of an indicator is litmus paper. Litmus paper is a way to measure whether a substance is acidic or basic. Blue litmus turns pink when an acid is placed on it and pink litmus turns blue when a base is placed on it. pH paper is a more accurate measure of pH, with the paper turning different colors depending on the pH value.

Spectrophotometry measures percent of light at different wavelengths absorbed and transmitted by a pigment solution.

Centrifugation involves spinning substances at a high speed. The more dense part of a solution will settle to the bottom of the test tube, where the lighter material will stay on top. Centrifugation is used to separate blood into blood cells and plasma, with the heavier blood cells settling to the bottom.

Electrophoresis uses electrical charges of molecules to separate them according to their size. The molecules, such as DNA or proteins are pulled through a gel towards either the positive end of the gel box (if the material has a negative charge) or the negative end of the gel box (if the material has a positive charge). DNA is negatively charged and moves towards the positive charge.

Use and care of living organisms in an ethical and appropriate manner

No dissections may be performed on living mammalian vertebrates or birds. Lower order life and invertebrates may be used. Biological experiments may be done with all animals except mammalian vertebrates or birds. No physiological harm may result to the animal. All animals housed and cared for in the school must be handled in a safe and humane manner. Animals are not to remain on school premises during extended vacations unless adequate care is provided. Any instructor who intentionally refuses to comply with the laws may be suspended or dismissed.

Pathogenic organisms must never be used for experimentation. Students should adhere to the following rules at all times when working with microorganisms to avoid accidental contamination:

1. Treat all microorganisms as if they were pathogenic.
2. Maintain sterile conditions at all times

SUBAREA VII. CHEMISTRY

Competency 0025 Understand the structure and nature of matter

Everything in our world is made up of **matter**, whether it is a rock, a building, an animal, or a person. Matter is defined by its characteristics: It takes up space and it has mass.

Mass is a measure of the amount of matter in an object. Two objects of equal mass will balance each other on a simple balance scale no matter where the scale is located. For instance, two rocks with the same amount of mass that are in balance on earth will also be in balance on the moon. They will feel heavier on earth than on the moon because of the gravitational pull of the earth. So, although the two rocks have the same mass, they will have different **weight.**

Weight is the measure of the earth's pull of gravity on an object. It can also be defined as the pull of gravity between other bodies. The units of weight measurement commonly used are the pound (English measure) and the kilogram (metric measure).

In addition to mass, matter also has the property of volume. **Volume** is the amount of cubic space that an object occupies. Volume and mass together give a more exact description of the object. Two objects may have the same volume, but different mass, or the same mass but different volumes, etc. For instance, consider two cubes that are each one cubic centimeter, one made from plastic, one from lead. They have the same volume, but the lead cube has more mass. The measure that we use to describe the cubes takes into consideration both the mass and the volume. **Density** is the mass of a substance contained per unit of volume. If the density of an object is less than the density of a liquid, the object will float in the liquid. If the object is denser than the liquid, then the object will sink.

Density is stated in grams per cubic centimeter (g/cm^3) where the gram is the standard unit of mass. To find an object's density, you must measure its mass and its volume. Then divide the mass by the volume ($D = m/V$).

To discover an object's density, first use a balance to find its mass. Then calculate its volume. If the object is a regular shape, you can find the volume by multiplying the length, width, and height together. However, if it is an irregular shape, you can find the volume by seeing how much water it displaces. Measure the water in the container before and after the object is submerged. The difference will be the volume of the object.

Specific gravity is the ratio of the density of a substance to the density of water. For instance, the specific density of one liter of turpentine is calculated by comparing its mass (0.81 kg) to the mass of one liter of water (1 kg):

$$\frac{\text{mass of 1 L alcohol}}{\text{mass of 1 L water}} = \frac{0.81 \text{ kg}}{1.00 \text{ kg}} = 0.81$$

Physical properties and chemical properties of matter describe the appearance or behavior of a substance. A **physical property** can be observed without changing the identity of a substance. For instance, you can describe the color, mass, shape, and volume of a book. **Chemical properties** describe the ability of a substance to be changed into new substances. Baking powder goes through a chemical change as it changes into carbon dioxide gas during the baking process.

An **element** is a substance that can not be broken down into other substances. To date, scientists have identified 109 elements: 89 are found in nature and 20 are synthetic.

An **atom** is the smallest particle of the element that retains the properties of that element. All of the atoms of a particular element are the same. The atoms of each element are different from the atoms of other elements.

Elements are assigned an identifying symbol of one or two letters. The symbol for oxygen is O and stands for one atom of oxygen. However, because oxygen atoms in nature are joined together in pairs, the symbol O_2 represents oxygen. This pair of oxygen atoms is a molecule. A **molecule** is an independent structural unit consisting of two or more atoms that are chemically bonded together. Molecules of oxygen, hydrogen, nitrogen, and chlorine consist of two atoms each.

A **compound** is made of two or more elements that have been chemically combined. Atoms join together when elements are chemically combined. The result is that the elements lose their individual identities when they are joined. The compound that they become has different properties.

We use a formula to show the elements of a chemical compound. A **chemical formula** is a shorthand way of showing what is in a compound by using symbols and subscripts. The letter symbols let us know what elements are involved and the number subscript tells how many atoms of each element are involved. No subscript is used if there is only one atom involved. For example, carbon dioxide is made up of one atom of carbon (C) and two atoms of oxygen (O_2), so the formula would be represented as CO_2.

Substances can combine without a chemical change. A **mixture** is any combination of two or more substances in which the substances keep their own properties. A fruit salad is a mixture. So is an ice cream sundae, although you might not recognize each part if it is stirred together. Colognes and perfumes are the other examples. You may not readily recognize the individual elements. However, they can be separated.

Compounds and **mixtures** are similar in that they are made up of two or more substances. However, they have the following opposite characteristics:

Compounds:
1. Made up of one kind of particle
2. Formed during a chemical change
3. Broken down only by chemical changes
4. Properties are different from its parts
5. Has a specific amount of each ingredient

Mixtures:
1. Made up of two or more particles
2. Not formed by a chemical change
3. Can be separated by physical changes
4. Properties are the same as its parts
5. Does not have a definite amount of each ingredient

Common compounds are **acids, bases, salts**, and **oxides** and are classified according to their characteristics.

An **acid** contains one element of hydrogen (H). Although it is never wise to taste a substance to identify it, acids have a sour taste. Vinegar and lemon juice are both acids, and acids occur in many foods in a weak state. Strong acids can burn skin and destroy materials. Common acids include:

Sulfuric acid (H_2SO_4)	-	Used in medicines, alcohol, dyes, and car batteries
Nitric acid (HNO_3)	-	Used in fertilizers, explosives, cleaning materials
Carbonic acid (H_2CO_3)	-	Used in soft drinks
Acetic acid ($HC_2H_3O_2$)	-	Used in making plastics, rubber, photographic film, and as a solvent

Bases have a bitter taste and the stronger ones feel slippery. Like acids, strong bases can be dangerous and should be handled carefully. All bases contain the elements oxygen and hydrogen (OH). Many household cleaning products contain bases. Common bases include:

Sodium hydroxide	$NaOH$	-	Used in making soap, paper, vegetable oils, and refining petroleum
Ammonium hydroxide	NH_4OH	-	Making deodorants, bleaching compounds, cleaning compounds
Potassium hydroxide	KOH	-	Making soaps, drugs, dyes, alkaline batteries, and purifying industrial gases
Calcium hydroxide	$Ca(OH)_2$	-	Making cement and plaster

An **indicator** is a substance that changes color when it comes in contact with an acid or a base. Litmus paper is an indicator. Blue litmus paper turns red in an acid. Red litmus paper turns blue in a base.

A substance that is neither acid nor base is **neutral**. Neutral substances do not change the color of litmus paper.

Salt is formed when an acid and a base combine chemically. Water is also formed. The process is called **neutralization**. Table salt (NaCl) is an example of this process. Salts are also used in toothpaste, Epsom salts, and cream of tartar. Calcium chloride ($CaCl_2$) is used on frozen streets and walkways to melt the ice.

Oxides are compounds that are formed when oxygen combines with another element. Rust is an oxide formed when oxygen combines with iron.

An **atom** is a nucleus surrounded by a cloud with moving electrons.

The **nucleus** is the center of the atom. The positive particles inside the nucleus are called **protons.** The mass of a proton is about 2,000 times that of the mass of an electron. The number of protons in the nucleus of an atom is called the **atomic number**. All atoms of the same element have the same atomic number.

Neutrons are another type of particle in the nucleus. Neutrons and protons have about the same mass, but neutrons have no charge. Neutrons were discovered because scientists observed that not all atoms in neon gas have the same mass. They had identified isotopes. **Isotopes** of an element have the same number of protons in the nucleus, but have different masses. Neutrons explain the difference in mass. They have mass but no charge.

The mass of matter is measured against a standard mass such as the gram. Scientists measure the mass of an atom by comparing it to that of a standard atom. The result is relative mass. The **relative mass** of an atom is its mass expressed in terms of the mass of the standard atom. The isotope of the element carbon is the standard atom. It has six (6) neutrons and is called carbon-12. It is assigned a mass of 12 atomic mass units (amu). Therefore, the **atomic mass unit (amu)** is the standard unit for measuring the mass of an atom. It is equal to the mass of a carbon atom.

The **mass number** of an atom is the sum of its protons and neutrons. In any element, there is a mixture of isotopes, some having slightly more or slightly fewer protons and neutrons. The **atomic mass** of an element is an average of the mass numbers of its atoms.

The following table summarizes the terms used to describe atomic nuclei:

Term	Example	Meaning	Characteristic
Atomic Number	# protons (p)	same for all atoms of a given element	Carbon (C) atomic number = 6 (6p)
Mass number	# protons + # neutrons (p + n)	changes for different isotopes of an element	C-12 (6p + 6n) C-13 (6p + 7n)
Atomic mass	average mass of the atoms of the element	usually not a whole number	atomic mass of carbon equals 12.011

Each atom has an equal number of electrons (negative) and protons (positive). Therefore, atoms are neutral. Electrons orbiting the nucleus occupy energy levels that are arranged in order and the electrons tend to occupy the lowest energy level available. A **stable electron arrangement** is an atom that has all of its electrons in the lowest possible energy levels.

Each energy level holds a maximum number of electrons. However, an atom with more than one level does not hold more than 8 electrons in its outermost shell.

Level	Name	Max. # of Electrons
First	K shell	2
Second	L shell	8
Third	M shell	18
Fourth	N shell	32

This can help explain why chemical reactions occur. Atoms react with each other when their outer levels are unfilled. When atoms either exchange or share electrons with each other, these energy levels become filled and the atom becomes more stable.

As an electron gains energy, it moves from one energy level to a higher energy level. The electron cannot leave one level until it has enough energy to reach the next level. **Excited electrons** are electrons that have absorbed energy and have moved farther from the nucleus.

Electrons can also lose energy. When they do, they fall to a lower level. However, they can only fall to the lowest level that has room for them. This explains why atoms do not collapse.

The **periodic table of elements** is an arrangement of the elements in rows and columns so that it is easy to locate elements with similar properties. The elements of the modern periodic table are arranged in numerical order by atomic number.

The **periods** are the rows down the left side of the table. They are called first period, second period, etc. The columns of the periodic table are called **groups**, or **families.** Elements in a family have similar properties.

There are three types of elements that are grouped by color: metals, nonmetals, and metalloids.

Element Key
Atomic
Number

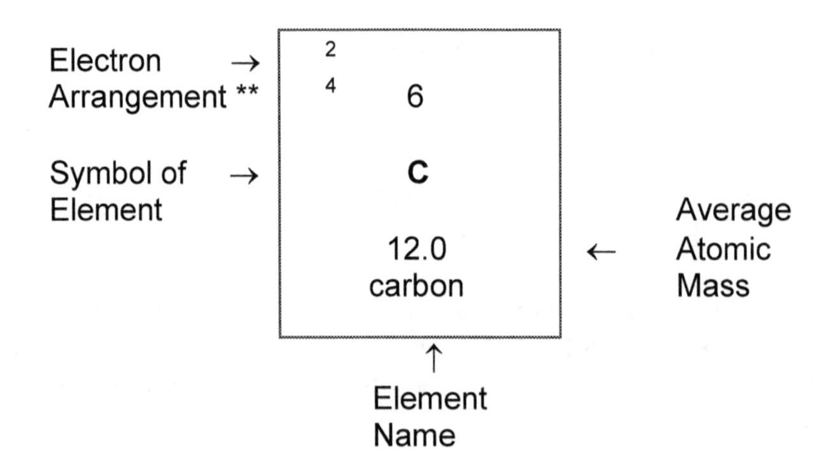

** Number of electrons on each level. Top number represents the innermost level.

The periodic table arranges metals into families with similar properties. The periodic table has its columns marked IA - VIIIA. These are the traditional group numbers. Arabic numbers 1 - 18 are also used, as suggested by the Union of Physicists and Chemists. The Arabic numerals will be used in this text.

Metals:

With the exception of hydrogen, all elements in Group 1 are **alkali metals**. These metals are shiny, softer and less dense than other metals, and are the most chemically active.

Group 2 metals are the **alkaline earth metals.** They are harder, denser, have higher melting points, and are chemically active.

The **transition elements** can be found by finding the periods (rows) from 4 to 7 under the groups (columns) 3 - 12. They are metals that do not show a range of properties as you move across the chart. They are hard and have high melting points. Compounds of these elements are colorful, such as silver, gold, and mercury.

Elements can be combined to make metallic objects. An **alloy** is a mixture of two or more elements having properties of metals. The elements do not have to be all metals. For instance, steel is made up of the metal iron and the non-metal carbon.

Nonmetals:

Nonmetals are not as easy to recognize as metals because they do not always share physical properties. However, in general the properties of nonmetals are the opposite of metals. They are dull, brittle, and are not good conductors of heat and electricity.

Nonmetals include solids, gases, and one liquid (bromine).

Nonmetals have four to eight electrons in their outermost energy levels and tend to attract electrons. As a result, the outer levels are usually filled with eight electrons. This difference in the number of electrons is what caused the differences between metals and nonmetals. The outstanding chemical property of nonmetals is that they react with metals.

The **halogens** can be found in Group 17. Halogens combine readily with metals to form salts. Table salt, fluoride toothpaste, and bleach all have an element from the halogen family.

The **Noble Gases** got their name from the fact that they did not react chemically with other elements, much like the nobility did not mix with the masses. These gases (found in Group 18) will only combine with other elements under very specific conditions. They are **inert** (inactive).

In recent years, scientists have found this to be only generally true, since chemists have been able to prepare compounds of krypton and xenon.

Metalloids:

Metalloids have properties in between metals and nonmetals. They can be found in Groups 13 - 16, but do not occupy the entire group. They are arranged in stair steps across the groups.

Physical Properties:

1. All are solids having the appearance of metals.
2. All are white or gray, but not shiny.
3. They will conduct electricity, but not as well as a metal.

Chemical Properties:

1. Have some characteristics of metals and nonmetals.
2. Properties do not follow patterns like metals and nonmetals. Each must be studied individually.

Boron is the first element in Group 13. It is a poor conductor of electricity at low temperatures. However, increase its temperature and it becomes a good conductor. By comparison, metals, which are good conductors, lose their ability as they are heated. It is because of this property that boron is so useful. Boron is a semiconductor. **Semiconductors** are used in electrical devices that have to function at temperatures too high for metals.

Silicon is the second element in Group 14. It is also a semiconductor and is found in great abundance in the earth's crust. Sand is made of a silicon compound, silicon dioxide. Silicon is also used in the manufacture of glass and cement.

Competency 0026 Understand the nature of physical changes in matter

The **phase of matter** (solid, liquid, or gas) is identified by its shape and volume. A **solid** has a definite shape and volume. A **liquid** has a definite volume, but no shape. A **gas** has no shape or volume because it will spread out to occupy the entire space of whatever container it is in.

Energy is the ability to cause change in matter. Applying heat to a frozen liquid changes it from solid back to liquid. Continue heating it and it will boil and give off steam, a gas.

Matter constantly changes. A **physical change** is a change that does not produce a new substance. The freezing and melting of water is an example of physical change. A **chemical change** (or chemical reaction) is any change of a substance into one or more other substances. Burning materials turn into smoke; a seltzer tablet fizzes into gas bubbles.

Evaporation is the change in phase from liquid to gas. **Condensation** is the change in phase from gas to liquid.

Competency 0027 Understand the nature of chemical changes in matter

One or more substances are formed during a **chemical reaction**. Also, energy is released during some chemical reactions. Sometimes the energy release is slow and sometimes it is rapid. In a fireworks display, energy is released very rapidly. However, the chemical reaction that produces tarnish on a silver spoon happens very slowly.

Chemical equilibrium is defined as occurring when the quantities of reactants and products are at a 'steady state' and no longer shifting, but the reaction may still proceed forward and backward. The rate of forward reaction must equal the rate of backward reaction.

In one kind of chemical reaction, two elements combine to form a new substance. We can represent the reaction and the results in a chemical equation.

Carbon and oxygen form carbon dioxide. The equation can be written:

C	+	O_2	→	CO_2
1 atom of carbon	+	2 atoms of oxygen	→ →	1 molecule of carbon dioxide

No matter is ever gained or lost during a chemical reaction; therefore the chemical equation must be *balanced.* This means that there must be the same number of atoms on both sides of the equation. Remember that the subscript numbers indicate the number of atoms in the elements. If there is no subscript, assume there is only one atom.

In a second kind of chemical reaction, the molecules of a substance split forming two or more new substances. An electric current can split water molecules into hydrogen and oxygen gas.

$2H_2O$	→	$2H_2$	+	O_2
2 molecules of water	→	2 molecules of hydrogen	+	1 molecule of oxygen

The number of molecules is shown by the number in front of an element or compound. If no number appears, assume that it is 1 molecule.

A third kind of chemical reaction is when elements change places with each other. An example of one element taking the place of another is when iron changes places with copper in the compound copper sulfate:

$CuSo_4$	+	Fe	→	$FeSO_4$	+	Cu
copper sulfate	+	iron (steel wool)		iron sulfate		copper

Sometimes two sets of elements change places. In this example, an acid and a base are combined:

HCl	+	NaOH	→	NaCl	+	H_2O
hydrochloric acid		sodium hydroxide		sodium chloride (table salt)		water

Matter can change, but it can not be created nor destroyed. The sample equations show two things:

1. In a chemical reaction, matter is changed into one or more different kinds of matter.
2. The amount of matter present before and after the chemical reaction is the same.

Many chemical reactions give off energy. Like matter, energy can change form but it can be neither created nor destroyed during a chemical reaction. This is the **law of conservation of energy.**

There are four kinds of chemical reactions:

In a **composition reaction**, two or more substances combine to form a compound.

$A + B \rightarrow AB$
i.e. silver and sulfur yield silver dioxide

In a **decomposition reaction**, a compound breaks down into two or more simpler substances.

$AB \rightarrow A + B$
i.e. water breaks down into hydrogen and oxygen

In a **single replacement reaction**, a free element replaces an element that is part of a compound.

A + BX $\rightarrow$ AX + B
i.e. iron plus copper sulfate yields iron sulfate plus copper

In a **double replacement reaction**, parts of two compounds replace each other. In this case, the compounds seem to switch partners.

AX + BY $\rightarrow$ AY + BX
i.e. sodium chloride plus mercury nitrate yields sodium nitrate plus mercury chloride

Competency 0028 Understand the kinetic molecular model of matter

As a substance is heated, the molecules begin moving faster within the container. As the substance becomes a gas and those molecules hit the sides of the container, pressure builds. **Pressure** is the force exerted on each unit of area of a surface. Pressure is measured in a unit called the **Pascal**. One Pascal (pa) is equal to one Newton of force pushing on one square meter of area.

Volume, temperature, and pressure of a gas are related.

Temperature and pressure: As the temperature of a gas increases, its pressure increases. When you drive a car, the friction between the road and the tire heats up the air inside the tire. Because the temperature increases, so does the pressure of the air on the inside of the tire.

Temperature and Volume: At a constant pressure, an increase in temperature causes an increase in the volume of a gas. If you apply heat to an enclosed container of gas, the pressure inside the bottle will increase as the heat increases. This is called **Charles' Law**.

These relations (pressure and temperature, and temperature and volume) are **direct variations**. As one component increases (decreases), the other also increases (decreases).

However, pressure and volume vary inversely.

Pressure and volume: At a constant temperature, a decrease in the volume of a gas causes an increase in its pressure. An example of this is a tire pump. The gas pressure inside the pump gets bigger as you press down on the pump handle because you are compressing the gas, or forcing it to exist in a smaller volume. This relationship between pressure and volume is called **Boyle's Law**.

SUBAREA VIII. PHYSICS

Competency 0029 Understand the concepts of force, motion, work, and power

Work and energy:

Work is done on an object when an applied force moves through a distance.

Power is the work done divided by the amount of time that it took to do it. (Power = Work / time)

Dynamics is the study of the relationship between motion and the forces affecting motion. **Force** causes motion.

Mass and weight are not the same quantities. An object's **mass** gives it a reluctance to change its current state of motion. It is also the measure of an object's resistance to acceleration. The force that the earth's gravity exerts on an object with a specific mass is called the object's weight on earth. Weight is a force that is measured in Newtons. Weight (W) = mass times acceleration due to gravity (**W = mg**). To illustrate the difference between mass and weight, picture two rocks of equal mass on a balance scale. If the scale is balanced in one place, it will be balanced everywhere, regardless of the gravitational field. However, the weight of the stones would vary on a spring scale, depending upon the gravitational field. In other words, the stones would be balanced both on earth and on the moon. However, the weight of the stones would be greater on earth than on the moon.

Newton's laws of motion:

Newton's first law of motion is also called the law of inertia. It states that an object at rest will remain at rest and an object in motion will remain in motion at a constant velocity unless acted upon by an external force.

Newton's second law of motion states that if a net force acts on an object, it will cause the acceleration of the object. The relationship between force and motion is Force equals mass times acceleration. **(F = ma).**

Newton's third law states that for every action there is an equal and opposite reaction. Therefore, if an object exerts a force on another object, that second object exerts an equal and opposite force on the first.

Surfaces that touch each other have a certain resistance to motion. This resistance is **friction.**

1. The materials that make up the surfaces will determine the magnitude of the frictional force.
2. The frictional force is independent of the area of contact between the two surfaces.
3. The direction of the frictional force is opposite to the direction of motion.
4. The frictional force is proportional to the normal force between the two surfaces in contact.

Static friction describes the force of friction of two surfaces that are in contact but do not have any motion relative to each other, such as a block sitting on an inclined plane. **Kinetic friction** describes the force of friction of two surfaces in contact with each other when there is relative motion between the surfaces.

When an object moves in a circular path, a force must be directed toward the center of the circle in order to keep the motion going. This constraining force is called **centripetal force**. Gravity is the centripetal force that keeps a satellite circling the earth.

Push and pull –Pushing a volleyball or pulling a bowstring applies muscular force when the muscles expand and contract. Elastic force is when any object returns to its original shape (for example, when a bow is released).

Rubbing – Friction opposes the motion of one surface past another. Friction is common when slowing down a car or sledding down a hill.

Pull of gravity – is a force of attraction between two objects. Gravity questions can be raised not only on earth but also between planets and even black hole discussions.

Forces on objects at rest – The formula F= ma is shorthand for force equals mass times acceleration. An object will not move unless the force is strong enough to move the mass. Also, there can be opposing forces holding the object in place. For instance, a boat may want to be forced by the currents to drift away but an equal and opposite force is a rope holding it to a dock.

Forces on a moving object - Overcoming inertia is the tendency of any object to oppose a change in motion. An object at rest tends to stay at rest. An object that is moving tends to keep moving.

Inertia and circular motion – The centripetal force is provided by the high banking of the curved road and by friction between the wheels and the road. This inward force that keeps an object moving in a circle is called centripetal force.

Competency 0030 Understand the concept of energy and the forms that energy can take

The relationship between heat, forms of energy, and work (mechanical, electrical, etc.) are the **Laws of Thermodynamics.** These laws deal strictly with systems in thermal equilibrium and not those within the process of rapid change or in a state of transition. Systems that are nearly always in a state of equilibrium are called **reversible systems.**

The first law of thermodynamics is a restatement of conservation of energy. The change in heat energy supplied to a system (Q) is equal to the sum of the change in the internal energy (U) and the change in the work done by the system against internal forces. $\Delta Q = \Delta U + \Delta W$

The second law of thermodynamics is stated in two parts:

1. No machine is 100% efficient. It is impossible to construct a machine that only absorbs heat from a heat source and performs an equal amount of work because some heat will always be lost to the environment.

2. Heat can not spontaneously pass from a colder to a hotter object. An ice cube sitting on a hot sidewalk will melt into a little puddle, but it will never spontaneously cool and form the same ice cube. Certain events have a preferred direction called the **arrow of time.**

Entropy is the measure of how much energy or heat is available for work. Work occurs only when heat is transferred from hot to cooler objects. Once this is done, no more work can be extracted. The energy is still being conserved, but is not available for work as long as the objects are the same temperature. Theory has it that, eventually, all things in the universe will reach the same temperature. If this happens, energy will no longer be usable.

Heat energy that is transferred into or out of a system is **heat transfer.** The temperature change is positive for a gain in heat energy and negative when heat is removed from the object or system.

The formula for heat transfer is $Q = mc\Delta T$ where Q is the amount of heat energy transferred, m is the amount of substance (in kilograms), c is the specific heat of the substance, and ΔT is the change in temperature of the substance. It is important to assume that the objects in thermal contact are isolated and insulated from their surroundings.

If a substance in a closed container loses heat, then another substance in the container must gain heat.

A **calorimeter** uses the transfer of heat from one substance to another to determine the specific heat of the substance.

When an object undergoes a change of phase it, goes from one physical state (solid, liquid, or gas) to another. For instance, water can go from liquid to solid (freezing) or from liquid to gas (boiling). The heat that is required to change from one state to the other is called **latent heat.**

The **heat of fusion** is the amount of heat that it takes to change from a solid to a liquid or the amount of heat released during the change from liquid to solid.

The **heat of vaporization** is the amount of heat that it takes to change from a liquid to a gaseous state.

Heat is transferred in three ways: **conduction, convection, and radiation.**

> **Conduction** occurs when heat travels through the heated solid.
>
> The transfer rate is the ratio of the amount of heat per amount of time it takes to transfer heat from area of an object to another. For example, if you place an iron pan on a flame, the handle will eventually become hot. How fast the handle gets too hot to handle is a function of the amount of heat and how long it is applied. Because the change in time is in the denominator of the function, the shorter the amount of time it takes to heat the handle, the greater the transfer rate.
>
> **Convection** is heat transported by the movement of a heated substance. Warmed air rising from a heat source such as a fire or electric heater is a common example of convection. Convection ovens make use of circulating air to more efficiently cook food.
>
> **Radiation** is heat transfer as the result of electromagnetic waves. The sun warms the earth by emitting radiant energy.

An example of all three methods of heat transfer occurs in the thermos bottle or Dewar flask. The bottle is constructed of double walls of Pyrex glass that have a space in between. Air is evacuated from the space between the walls and the inner wall is silvered. The lack of air between the walls lessens heat loss by convection and conduction. The heat inside is reflected by the silver, cutting down heat transfer by radiation. Hot liquids remain hotter and cold liquids remain colder for longer periods of time.

Competency 0031 Understand characteristics of waves and the behavior of sound and light waves

Sound waves are produced by a vibrating body. The vibrating object moves forward and compresses the air in front of it, then reverses direction so that the pressure on the air is lessened and expansion of the air molecules occurs. One compression and expansion creates one longitudinal wave. Sound can be transmitted through any gas, liquid, or solid. However, it cannot be transmitted through a vacuum, because there are no particles present to vibrate and bump into their adjacent particles to transmit the wave.

The vibrating air molecules move back and forth parallel to the direction of motion of the wave as they pass the energy from adjacent air molecules (closer to the source) to air molecules farther away from the source.

The **pitch** of a sound depends on the **frequency** that the ear receives. High-pitched sound waves have high frequencies. High notes are produced by an object that is vibrating at a greater number of times per second than one that produces a low note.

The **intensity** of a sound is the amount of energy that crosses a unit of area in a given unit of time. The loudness of the sound is subjective and depends upon the effect on the human ear. Two tones of the same intensity but different pitches may appear to have different loudness. The intensity level of sound is measured in decibels. Normal conversation is about 60 decibels. A power saw is about 110 decibels.

The **amplitude** of a sound wave determines its loudness. Loud sound waves have large amplitudes. The larger the sound wave, the more energy is needed to create the wave.

An oscilloscope is useful in studying waves because it gives a picture of the wave that shows the crest and trough of the wave. **Interference** is the interaction of two or more waves that meet. If the waves interfere constructively, the crest of each one meets the crests of the others. They combine into a crest with greater amplitude. As a result, you hear a louder sound. If the waves interfere destructively, then the crest of one meets the trough of another. They produce a wave with lower amplitude that produces a softer sound.

If you have two tuning forks that produce different pitches, then one will produce sounds of a slightly higher frequency. When you strike the two forks simultaneously, you may hear beats. **Beats** are a series of loud and soft sounds. This is because when the waves meet, the crests combine at some points and produce loud sounds. At other points, they nearly cancel each other out and produce soft sounds.

Shadows illustrate one of the basic properties of light. Light travels in a straight line. If you put your hand between a light source and a wall, you will interrupt the light and produce a shadow.

When light hits a surface, it is **reflected.** The angle of the incoming light (angle of incidence) is the same as the angle of the reflected light (angle of reflection). It is this reflected light that allows you to see objects. You see the objects when the reflected light reaches your eyes.

Different surfaces reflect light differently. Rough surfaces scatter light in many different directions. A smooth surface reflects the light in one direction. If it is smooth and shiny (like a mirror), you see your image in the surface.

When light enters a different medium, it bends. This bending, or change of speed, is called **refraction**.

Light can be **diffracted**, or bent around the edges of an object. Diffraction occurs when light goes through a narrow slit. As light passes through it, the light bends slightly around the edges of the slit. You can demonstrate this by pressing your thumb and forefinger together, making a very thin slit between them. Hold them about 8 cm from your eye and look at a distant source of light. The pattern you observe is caused by the diffraction of light.

Light and other electromagnetic radiation can be polarized because the waves are transverse. The distinguishing characteristic of transverse waves is that they are perpendicular to the direction of the motion of the wave. Polarized light has vibrations confined to a single plane that is perpendicular to the direction of motion. Light is able to be polarized by passing it through special filters that block all vibrations except those in a single plane. By blocking out all but one place of vibration, polarized sunglasses cut down on glare.

Light can travel through thin fibers of glass or plastic without escaping the sides. Light on the inside of these fibers is reflected so that it stays inside the fiber until it reaches the other end. Such fiber optics are being used to carry telephone messages. Sound waves are converted to electric signals which are coded into a series of light pulses which move through the optical fiber until they reach the other end. At that time, they are converted back into sound.

The image that you see in a bathroom mirror is a virtual image because it only seems to be where it is. However, a curved mirror can produce a real image. A real image is produced when light passes through the point where the image appears. A real image can be projected onto a screen.

Cameras use a convex lens to produce an image on the film. A **convex lens** is thicker in the middle than at the edges. The image size depends upon the focal length (distance from the focus to the lens). The longer the focal length, the larger is the image. A **converging lens** produces a real image whenever the object is far enough from the lens so that the rays of light from the object can hit the lens and be focused into a real image on the other side of the lens.

Eyeglasses can help correct deficiencies of sight by changing where the image seen is focused on the retina of the eye. If a person is nearsighted, the lens of his eye focuses images in front of the retina. In this case, the corrective lens placed in the eyeglasses will be concave so that the image will reach the retina. In the case of farsightedness, the lens of the eye focuses the image behind the retina. The correction will call for a convex lens to be fitted into the glass frames so that the image is brought forward into sharper focus.

Competency 0032 Understand principles of electricity, magnetism, and electromagnetism

The electromagnetic spectrum is measured in frequency (f) in hertz and wavelength (λ) in meters. The frequency times the wavelength of every electromagnetic wave equals the speed of light (3.0 x 10^9 meters/second).

Roughly, the range of wavelengths of the electromagnetic spectrum is:

	f	λ
Radio waves	10^{5}- 10^{-1} hertz	10^{3} -10^{9} meters
Microwaves	10^{-1} - 10^{-3} hertz	10^{9} -10^{11} meters
Infrared radiation	10^{-3} - 10^{-6} hertz	$10^{11.2}$-$10^{14.3}$ meters
Visible light	$10^{-6.2}$ $10^{-6.9}$ hertz	$10^{14.3}$-10^{15} meters
Ultraviolet radiation	10^{-7} - 10^{-9} hertz	10^{15} -$10^{17.2}$ meters
X-Rays	10^{-9} - 10^{-11} hertz	$10^{17.2}$ -10^{19} meters
Gamma Rays	10^{-11}- 10^{-15} hertz	10^{19} - $10^{23.25}$meters

Electrostatics is the study of stationary electric charges. A plastic rod that is rubbed with fur or a glass rod that is rubbed with silk will become electrically charged and will attract small pieces of paper. The charge on the plastic rod rubbed with fur is negative and the charge on glass rod rubbed with silk is positive.

Electrically charged objects share these characteristics:

1. Like charges repel one another.
2. Opposite charges attract each other.
3. Charge is conserved. A neutral object has no net change. If the plastic rod and fur are initially neutral, when the rod becomes charged by the fur, a negative charge is transferred from the fur to the rod. The net negative charge on the rod is equal to the net positive charge on the fur.

Materials through which electric charges can easily flow are called conductors. On the other hand, an **insulator** is a material through which electric charges do not move easily, if at all. A simple device used to indicate the existence of a positive or negative charge is called an **electroscope**. An electroscope is made up of a conducting knob and attached to it are very lightweight conducting leaves usually made of foil (gold or aluminum). When a charged object touches the knob, the leaves push away from each other because like charges repel. It is not possible to tell whether the charge is positive or negative.

Charging by induction:

Touch the knob with a finger while a charged rod is nearby. The electrons will be repulsed and flow out of the electroscope through the hand. If the hand is removed while the charged rod remains close, the electroscope will retain the charge.

When an object is rubbed with a charged rod, the object will take on the same charge as the rod. However, charging by induction gives the object the opposite charge as that of the charged rod.

Grounding charge:

Charge can be removed from an object by connecting it to the earth through a conductor. The removal of static electricity by conduction is called **grounding**.

An **electric circuit** is a path along which electrons flow. A simple circuit can be created with a dry cell, wire, a bell, or a light bulb. When all are connected, the electrons flow from the negative terminal, through the wire to the device and back to the positive terminal of the dry cell. If there are no breaks in the circuit, the device will work. The circuit is closed. Any break in the flow will create an open circuit and cause the device to shut off.

The device (bell, bulb) is an example of a **load**. A load is a device that uses energy. Suppose that you also add a buzzer so that the bell rings when you press the buzzer button. The buzzer is acting as a **switch**. A switch is a device that opens or closes a circuit. Pressing the buzzer makes the connection complete and the bell rings. When the buzzer is not engaged, the circuit is open and the bell is silent.

A **series circuit** is one where the electrons have only one path along which they can move. When one load in a series circuit goes out, the circuit is open. An example of this is a set of Christmas tree lights that is missing a bulb. None of the bulbs will work.

A **parallel circuit** is one where the electrons have more than one path to move along. If a load goes out in a parallel circuit, the other load will still work because the electrons can still find a way to continue moving along the path.

When an electron goes through a load, it does work and therefore loses some of its energy. The measure of how much energy is lost is called the **potential difference**. The potential difference between two points is the work needed to move a charge from one point to another.

Potential difference is measured in a unit called the volt. **Voltage** is potential difference. The higher the voltage, the more energy the electrons have. This energy is measured by a device called a voltmeter. To use a voltmeter, place it in a circuit parallel with the load you are measuring.

Current is the number of electrons per second that flow past a point in a circuit. Current is measured with a device called an ammeter. To use an ammeter, put it in series with the load you are measuring.

As electrons flow through a wire, they lose potential energy. Some is changed into heat energy because of resistance. **Resistance** is the ability of the material to oppose the flow of electrons through it. All substances have some resistance, even if they are a good conductor such as copper. This resistance is measured in units called **ohms**. A thin wire will have more resistance than a thick one because it will have less room for electrons to travel. In a thicker wire, there will be more possible paths for the electrons to flow. Resistance also depends upon the length of the wire. The longer the wire, the more resistance it will have. Potential difference, resistance, and current form a relationship know as **Ohm's Law**. Current **(I)** is measured in amperes and is equal to potential difference **(V)** divided by resistance **(R)**. Current = Potential Dif. / resistance

$$I = V / R$$

If you have a wire with resistance of 5 ohms and a potential difference of 75 volts, you can calculate the current by

I = 75 volts / 5 ohms
I = 15 amperes

A current of 10 or more amperes will cause a wire to get hot. 22 amperes is about the maximum for a house circuit. Anything above 25 amperes can start a fire.

Electricity can be used to change the chemical composition of a material. For instance, when electricity is passed through water, it breaks the water down into hydrogen gas and oxygen gas.

Circuit breakers in a home monitor the electric current. If there is an overload, the circuit breaker will create an open circuit, stopping the flow of electricity.

Computers can be made small enough to fit inside a plastic credit card by creating what is known as a solid state device. In this device, electrons flow through solid material such as silicon.

Resistors are used to regulate volume on a television or radio or through a dimmer switch for lights.

A bird can sit on an electrical wire without being electrocuted because the bird and the wire have about the same potential. However, if that same bird would touch two wires at the same time he would not have to worry about flying south next year.

When caught in an electrical storm, a car is a relatively safe place from lightening because of the resistance of the rubber tires. A metal building would not be safe unless there was a lightening rod that would attract the lightening and conduct it into the ground.

Magnets

Magnets have a north pole and a south pole. Like poles repel and opposing poles attract. A **magnetic field** is the space around a magnet where its force will affect objects. The closer you are to a magnet, the stronger the force. As you move away, the force becomes weaker.

Some materials act as magnets and some do not. This is because magnetism is a result of electrons in motion. The most important motion in this case is the spinning of the individual electrons. Electrons spin in pairs in opposite directions in most atoms. Each spinning electron has the magnetic field that it creates canceled out by the electron that is spinning in the opposite direction.

In an atom of iron, there are four unpaired electrons. The magnetic fields of these are not canceled out. Their fields add up to make a tiny magnet. There fields exert forces on each other setting up small areas in the iron called **magnetic domains** where atomic magnetic fields line up in the same direction.

You can make a magnet out of an iron nail by stroking the nail in the same direction repeatedly with a magnet. This causes poles in the atomic magnets in the nail to be attracted to the magnet. The tiny magnetic fields in the nail line up in the direction of the magnet. The magnet causes the domains pointing in its direction to grow in the nail. Eventually, one large domain results and the nail becomes a magnet.

A bar magnet has a north pole and a south pole. If you break the magnet in half, each piece will have a north and south pole.

The earth has a magnetic field. In a compass, a tiny, lightweight magnet is suspended and will line its south pole up with the North Pole magnet of the earth.

A magnet can be made out of a coil of wire by connecting the ends of the coil to a battery. When the current goes through the wire, the wire acts in the same way that a magnet does, it is called an **electromagnet**. The poles of the electromagnet will depend upon which way the electric current runs. An electromagnet can be made more powerful in three ways:

1. Make more coils.
2. Put an iron core (nail) inside the coils.
3. Use more battery power.

SUBAREA IX. BIOLOGY

Competency 0033 Understand the characteristics and life processes of living organisms

The organization of living systems builds by levels from small to increasingly more large and complex. All aspects, whether it be a cell or an ecosystem, have the same requirements to sustain life. Life is organized from simple to complex in the following way:

> **Organelles** make up **cells,** which make up **tissues,** which make up **organs**. Groups of organs make up **organ systems**. Organ systems work together to provide life for the **organism.**

Several characteristics have been described to identify living versus non-living substances.

1. **Living things are made of cells**; they grow, are capable of reproduction and respond to stimuli.
2. **Living things must adapt to environmental changes or perish**.
3. **Living things carry on metabolic processes**. They use and make energy.

The structure of the cell is often related to the cell's function. Root hair cells differ from flower stamens or leaf epidermal cells. They all have different functions.

Animal cells – begin a discussion of the nucleus as a round body inside the cell. It controls the cell's activities. The nuclear membrane (nuclear envelope) contains threadlike structures called chromosomes which contain genes. The genes are units that control cell activities found in the nucleus. The cytoplasm has many structures in it. Vacuoles contain the food for the cell. Other vacuoles contain waste materials. Animal cells differ from plant cells because they have cell membranes.

Plant cells – have cell walls. A cell wall differs from cell membranes. The cell membrane is very thin and is a part of the cell. The cell wall is thick and is a nonliving part of the cell. Chloroplasts are bundles of chlorophyll.

Single cells – A single celled organism is called a **protist.** When you look under a microscope the animal-like protists are called **protozoans.** They do not have chloroplasts. They are usually classified by the way they move for food. Amoebas engulf other protists by flowing around and over them. The paramecium has a hair-like structure that allows it to move back and forth like tiny oars searching for food. The euglena is an example of a protozoan that moves with a tail-like structure called a flagellum.

Plant-like protists have cell walls and float in the ocean. **Bacteria** are the simplest protists. A bacterial cell is surrounded by a cell wall but there is no nucleus inside the cell. Most bacteria do not contain chlorophyll so they do not make their own food. The classification of bacteria is by shape. Cocci are round, bacilli are rod-shaped, and spirilla are spiral-shaped.

Prokaryotic cells consist only of bacteria and blue-green algae. Bacteria were most likely the first cells and date back in the fossil record to 3.5 billion years ago. The important things that put these cells in their own group are:

1. They have no defined nucleus or nuclear membrane. The DNA and ribosomes float freely within the cell.

2. They have a thick cell wall. This is for protection, to give shape, and to keep the cell from bursting.

3. The cell walls contain amino sugars (glycoproteins). Penicillin works by disrupting the cell wall, which is bad for the bacteria but will not harm the host.

4. Some have a capsule made of polysaccharides which make the bacteria sticky.

5. Some have pili, which is a protein strand. This also allows for attachment of the bacteria and may be used for sexual reproduction (conjugation).

6. Some have flagella for movement.

Eukaryotic cells are found in protists, fungi, plants and animals. Some features of eukaryotic cells include:

1. They have a nucleus and chromosomes located in the nucleus.

2. They are usually larger than prokaryotic cells.

3. They contain many organelles, which are membrane-bound areas for specific cell functions.

4. They contain a cytoskeleton which provides a protein framework for the cell.

5. They contain cytoplasm to support the organelles and contain the ions and molecules necessary for cell function.

Parts of Eukaryotic Cells

1. **Nucleus** - The brain of the cell. The nucleus contains:

chromosomes- DNA, RNA and proteins tightly coiled to conserve space while providing a large surface area.
chromatin - loose structure of chromosomes. Chromosomes are called chromatin when the cell is not dividing.
nucleoli - where ribosomes are made. These are seen as dark spots in the nucleus.
nuclear membrane - contains pores which let RNA out of the nucleus. The nuclear membrane is continuous with the endoplasmic reticulum which allows the membrane to expand or shrink if needed.

2. **Ribosomes** - the site of protein synthesis. Ribosomes may be free floating in the cytoplasm or attached to the endoplasmic reticulum. There may be up to a half a million ribosomes in a cell, depending on how much protein is made by the cell.

3. **Endoplasmic Reticulum** - These are folded and provide a large surface area. They are the "roadway" of the cell and allow for the transport of materials. The lumen of the endoplasmic reticulum helps to keep materials out of the cytoplasm and headed in the right direction. The endoplasmic reticulum is capable of building new membrane material. There are two types:

Smooth Endoplasmic Reticulum - contain no ribosomes on their surface.

Rough Endoplasmic Reticulum - contain ribosomes on their surface. This form of ER is abundant in cells that make many proteins, like in the pancreas, which produces many digestive enzymes.

4. **Golgi Complex or Golgi Apparatus** - This structure is stacked to increase surface area. The Golgi Complex functions to sort, modify and package molecules that are made in other parts of the cell. These molecules are either sent out of the cell or to other organelles within the cell.

5. **Lysosomes** - found mainly in animal cells. These contain digestive enzymes that break down food, substances not needed, viruses, damaged cell components, and eventually the cell itself. It is believed that lysosomes are responsible for the aging process.

6. Mitochondria - large organelles that make ATP to supply energy to the cell. Muscle cells have many mitochondria because they use a great deal of energy. The folds inside the mitochondria are called cristae. They provide a large surface where the reactions of cellular respiration occur. Mitochondria have their own DNA and are capable of reproducing themselves if a greater demand is made for additional energy. Mitochondria are found only in animal cells.

7. Plastids - found in photosynthetic organisms only. They are similar to the mitochondria due to their double membrane structure. They also have their own DNA and can reproduce if increased capture of sunlight becomes necessary. There are several types of plastids: Plants!

Chloroplasts - green; function in photosynthesis. They are capable of trapping sunlight.
Chromoplasts - make and store yellow and orange pigments; they provide color to leaves, flowers and fruits.
Amyloplasts - store starch and are used as a food reserve. They are abundant in roots like potatoes.

8. Cell Wall - Found in plant cells only, it is composed of cellulose and fibers. It is thick enough for support and protection, yet porous enough to allow water and dissolved substances to enter. Cell walls are cemented to each other.

9. Vacuoles - hold stored food and pigments. Vacuoles are very large in plants. This is allows them to fill with water in order to provide turgor pressure. Lack of turgor pressure causes a plant to wilt.

10. Cytoskeleton - composed of protein filaments attached to the plasma membrane and organelles. They provide a framework for the cell and aid in cell movement. They constantly change shape and move about. Three types of fibers make up the cytoskeleton:

Microtubules - largest of the three; make up cilia and flagella for locomotion. Flagella grow from a basal body. Some examples are sperm cells and tracheal cilia. Centrioles are also composed of microtubules. They form the spindle fibers that pull the cell apart into two cells during cell division. Centrioles are not found in the cells of higher plants.

Intermediate Filaments - they are smaller than microtubules but larger than microfilaments. They help the cell to keep its shape.

Microfilaments - smallest of the three, they are made of actin and small amounts of myosin (like in muscle cells). They function in cell movement such as cytoplasmic streaming, endocytosis, and ameboid movement. Microfilaments pinch the two cells apart after cell division, forming two cells.

The purpose of cell division is to provide growth and repair in body (somatic) cells and to replenish or create sex cells for reproduction. There are two forms of cell division. Mitosis is the division of somatic cells and **meiosis** is the division of sex cells (eggs and sperm). The table below summarizes the major differences between the two processes.

MITOSIS	**MEIOSIS**
1. Division of somatic cell	1. Division of sex cells
2. Two cells result from each division	2. Four cells or polar bodies result from each division
3. Chromosome number is identical to parent cells	3. Chromosome number is half the number of parent cells
4. For cell growth and repair	4. Recombinations provide genetic diversity

Some terms to know:

gamete - sex cell or germ cell; eggs and sperm.
chromatin - loose chromosomes; this state is found when the cell is not dividing.
chromosome - tightly coiled, visible chromatin; this state is found when the cell is dividing.
homologues - chromosomes that contain the same information. They are of the same length and contain the same genes.
diploid - 2n number; diploid chromosomes are a pair of chromosomes (somatic cells).
haploid - 1n number; haploid chromosomes are a half of a pair (sex cells).

MITOSIS

The cell cycle is the life cycle of the cell. It is divided into two stages; **Interphase** and the **mitotic division** where the cell is actively dividing. Interphase is divided into three steps; G1 (growth) period, where the cell is growing and metabolizing, S period (synthesis) where new DNA and enzymes are being made and the G2 phase (growth) where new proteins and organelles are being made to prepare for cell division. The mitotic stage consists of the stages of mitosis and the division of the cytoplasm.

The stages of mitosis and their events are as follows. Be sure to know the correct order of steps. (IPMAT)

1. **Interphase** - chromatin is loose, chromosomes are replicated, cell metabolism is occurring. Interphase is technically not a stage of mitosis.

2. **Prophase** - once the cell enters prophase, it proceeds through the following steps continuously with no stopping. The chromatin condenses to become visible chromosomes. The nucleolus disappears and the nuclear membrane (envelope) breaks apart. Mitotic spindles form which will eventually pull the chromosomes apart. They are composed of microtubules. The cytoskeleton breaks down and the spindles are pushed to the poles or opposite ends of the cell by the action of centrioles.

3. **Metaphase** - kinetechore fibers attach to the chromosomes which causes the chromosomes to line up in the center of the cell (think **m**iddle for **m**etaphase)

4. **Anaphase** - centromeres split in half and homologous chromosomes separate. The chromosomes are pulled to the poles of the cell, with identical sets at either end.

5. **Telophase** - two nuclei form with a full set of DNA that is identical to the parent cell. The nucleoli become visible and the nuclear membrane reassembles. A cell plate is visible in plant cells, whereas a cleavage furrow is formed in animal cells. The cell is pinched into two cells. Cytokinesis, or division, of the cytoplasm and organelles occurs.

Meiosis contains the same five stages as mitosis, but is repeated in order to reduce the chromosome number by one half. This way, when the sperm and egg join during fertilization, the haploid number is reached. The steps of meiosis are as follows:

Major function of Meiosis I - chromosomes are replicated; cells remain diploid.

Prophase I - replicated chromosomes condense and pair with homologues. This forms a tetrad. Crossing over (the exchange of genetic material between homologues to further increase diversity) occurs during Prophase I.
Metaphase I - homologous sets attach to spindle fibers after lining up in the middle of the cell.
Anaphase I - sister chromatids remain joined and move to the poles of the cell.
Telophase I - two new cells are formed; chromosome number is still diploid

Major function of Meiosis II - to reduce the chromosome number in half.

Prophase II - chromosomes condense.
Metaphase II - spindle fibers form again, sister chromatids line up in center of cell, centromeres divide and sister chromatids separate.
Anaphase II - separated chromosomes move to opposite ends of the cell.
Telophase II - four haploid cells form for each original sperm germ cell. One viable egg cell gets all the genetic information and three polar bodies form with no DNA. The nuclear membrane reforms and cytokinesis occurs.

DNA and DNA REPLICATION

The modern definition of a gene is a unit of genetic information. DNA makes up genes which. in turn, make up the chromosomes. DNA is wound tightly around proteins in order to conserve space. The DNA/protein combination makes up the chromosome. DNA controls the synthesis of proteins, thereby controlling the total cell activity. DNA is capable of making copies of itself.

Review of DNA structure:

1. Made of nucleotides; a five carbon sugar, phosphate group and nitrogen base (either adenine, guanine, cytosine or thymine).

2. Consists of a sugar/phosphate backbone which is covalently bonded. The bases are joined down the center of the molecule and are attached by hydrogen bonds which are easily broken during replication.

3. The amount of adenine equals the amount of thymine and the amount of cytosine equals the amount of guanine.

4. The shape is that of a twisted ladder called a double helix. The sugar/phosphates make up the sides of the ladder and the base pairs make up the rungs of the ladder.

DNA Replication

Enzymes control each step of the replication of DNA. The molecule untwists. The hydrogen bonds between the bases break and serve as a pattern for replication. Free nucleotides found inside the nucleus join to form a new strand. Two new pieces of DNA are formed which are identical. This is a very accurate process. There is only one mistake for every billion nucleotides added. This is because there are enzymes (polymerases) present that proofread the molecule. In eukaryotes, replication occurs in many places along the DNA at once. The molecule may open up at many places like a broken zipper. In prokaryotic circular plasmids, replication begins at a point on the plasmid and goes in both directions until it meets itself.

Base pairing rules are important in determining a new strand of DNA sequence. For example, say our original strand of DNA had the sequence as follows:

1. A T C G G C A A T A G C This may be called our sense strand as it contains a sequence that makes sense or codes for something. The complementary strand (or other side of the ladder) would follow base pairing rules (A bonds with T and C bonds with G) and would read:

2. T A G C C G T T A T C G When the molecule opens up and nucleotides add on, the base pairing rules create two new identical strands of DNA

1. A T C G G C A A T A G C and 2. A T C G G C A A T A G C
T A G C C G T T A T C G T A G C C G T T A T C G

Protein Synthesis

It is necessary for cells to manufacture new proteins for growth and repair of the organism. Protein synthesis is the process that allows the DNA code to be read and carried out of the nucleus into the cytoplasm in the form of RNA. This is where the ribosomes are found, which are the sites of protein synthesis. The protein is then assembled according to the instructions on the DNA. There are several types of RNA. Familiarize yourself with where they are found and their function.

Messenger RNA - (mRNA) copies the code from DNA in the nucleus and takes it to the ribosomes in the cytoplasm.

Transfer RNA - (tRNA) is free floating in the cytoplasm. Its job is to carry and position amino acids for assembly on the ribosome.

Ribosomal RNA - (rRNA) found in the ribosomes. They make a place for the proteins to be made. rRNA is believed to have many important functions; so much research is currently being done currently in this area.

Along with enzymes and amino acids, the RNA's function is to assist in the building of proteins. There are two stages of protein synthesis:

Transcription - this phase allows for the assembly of mRNA and occurs in the nucleus where the DNA is found. The DNA splits open and the RNA polymerase reads the DNA sequence and "transcribes" a sequence of single stranded mRNA. For example, if the code on the DNA is T A C C T C G T A C G A, the mRNA will make a complementary strand reading: A U G G A G C A U G C U (Remember that uracil replaces thymine in RNA.) The mRNA leaves the nucleus and enters the cytoplasm.

Translation – In the cytoplasm, the mRNA binds to the ribosome. The ribosome is made of rRNA (ribosomal RNA) and protein. The ribosome "reads" three nucleotides of the mRNA at a time. Groups of three mRNA nucleotides are called codons. The codon will eventually code for a specific amino acid to be carried to the ribosome. "Start" codons begin the building of the protein and "stop" codons end translation. When the stop codon is reached, the ribosome stops reading the mRNA and releases the protein and mRNA. The nucleotide sequence is translated to choose the correct amino acid sequence. As the rRNA translates the code at the ribosome, tRNAs which contain an **anticodon** seek out the correct amino acid and bring it back to the ribosome. For example, using the codon sequence from the example above:

the mRNA reads A U G / G A G / C A U / G C U
the anticodons are U A C / C U C / G U A / C G A
the amino acid sequence would be: Methionine (start) - Glu - His - Ala.

*Be sure to note if the table you are given is written according to the codon sequence or the anticodon sequence. It will be specified.

This whole process is accomplished through the assistance of **activating enzymes**. Each of the twenty amino acids has their own enzyme. The enzyme binds the amino acid to the tRNA. When the amino acids get close to each other on the ribosome, they bond together using peptide bonds. The start and stop codons are called nonsense codons. There is one start codon (AUG) and three stop codons. (UAA, UGA and UAG). Addition mutations will cause the whole code to shift, thereby producing the wrong protein or, at times, no protein at all.

Photosynthesis is the process by which plants make carbohydrates from the energy of the sun, carbon dioxide, and water. Oxygen is a waste product. Photosynthesis occurs in the chloroplast where the pigment chlorophyll traps sun energy. It is divided into two major steps:

Light Reactions - Sunlight is trapped, water is split, and oxygen is given off. ATP is made and hydrogens reduce NADP to $NADPH_2$. The light reactions occur in light. The products of the light reactions enter into the dark reactions (Calvin cycle).
Dark Reactions - Carbon dioxide enters during the dark reactions which can occur with or without the presence of light. The energy transferred from $NADPH_2$ and ATP allow for the fixation of carbon into glucose.

Respiration - during times of decreased light, plants break down the products of photosynthesis through cellular respiration. Glucose, with the help of oxygen, breaks down and produces carbon dioxide and water as waste. Approximately fifty percent of the products of photosynthesis are used by the plant for energy.

Transpiration - water travels up the xylem of the plant through the process of transpiration. Water sticks to itself (cohesion) and to the walls of the xylem (adhesion). As it evaporates through the stomata of the leaves, the water is pulled up the column from the roots. Environmental factors such as heat and wind increase the rate of transpiration. High humidity will decrease the rate of transpiration.

Competency 0034 Understand principles related to the inheritance of characteristics

Gregor Mendel is recognized as the father of genetics. His work in the late 1800's is the basis of our knowledge of genetics. Although unaware of the presence of DNA or genes, Mendel realized there were factors (now known as genes) that were transferred from parents to their offspring. Mendel worked with pea plants and fertilized the plants himself, keeping track of subsequent generations which led to the Mendelian laws of genetics. Mendel found that two "factors" governed each trait, one from each parent. Traits or characteristics came in several forms, known as alleles. For example, the trait of flower color had white alleles and purple alleles. Mendel formed three laws:

Law of dominance - in a pair of alleles, one trait may cover up the allele of the other trait. Example: brown eyes are dominant to blue eyes.

Law of segregation - only one of the two possible alleles from each parent is passed on to the offspring from each parent. (During meiosis, the haploid number insures that half the sex cells get one allele, half get the other).

Law of independent assortment - alleles sort independently of each other. (Many combinations are possible depending on which sperm ends up with which egg. Compare this to the many combinations of hands possible when dealing a deck of cards).

monohybrid cross - a cross using only one trait.

dihybrid cross - a cross using two traits. More combinations are possible.

Punnet squares - these are used to show the possible ways that genes combine and indicate probability of the occurrence of a certain genotype or phenotype. One parent's genes are put at the top of the box and the other parent at the side of the box. Genes combine on the square just like numbers that are added in addition tables we learned in elementary school.

Example: Monohybrid Cross - four possible gene combinations

	T	t
T	TT	Tt
t	Tt	tt

Example: Dihybrid Cross - sixteen possible gene combinations

R = roundness
r = wrinkled

Y= yellow
y = green

	RY	Ry	rY	ry
RY	RRYY	RRYy	RrYY	RrYy
Ry	RRYy	RRyy	RrYy	Rryy
rY	RrYY	RrYy	rrYY	rrYy
ry	RrYy	Rryy	rrYy	rryy

Since it's not a perfect world, mistakes happen. Inheritable changes in DNA are called **mutations**. Mutations may be errors in replication or a spontaneous rearrangement of one or more segments by factors like radioactivity, drugs, or chemicals. The amount of the change is not as critical as where the change is. Mutations may occur in somatic or sex cells. Usually the ones on sex cells are more dangerous since they contain the basis of all information for the developing offspring. Mutations are not always bad. They are the basis of evolution and, if they make a more favorable variation that enhances the organism's survival, then they are beneficial. But, mutations may also lead to abnormalities, birth defects, and even death. There are several types of mutations; let's suppose a normal sequence was as follows:

Normal - A B C D E F

Duplication - one gene is repeated. A B C C D E F

Inversion - a segment of the sequence is flipped around. A E D C B F

Deletion - a gene is left out. A B C E F

Insertion or Translocation - a segment from another place on the DNA is inserted in the wrong place. A B C R S D E F

Breakage - a piece is lost. A B C (DEF is lost)

Nondisjunction – This occurs during meiosis when chromosomes fail to separate properly. One sex cell may get both genes and another may get none. Depending on the chromosomes involved, this may or may not be serious. Offspring end up with either an extra chromosome or are missing one. An example of nondisjunction is Down syndrome, where three of chromosome #21 are present.

SOME DEFINITIONS TO KNOW

Dominant - the stronger of the two traits. If a dominant gene is present, it will be expressed. Shown by a capital letter.

Recessive - the weaker of the two traits. In order for the recessive gene to be expressed, there must be two recessive genes present. Shown by a lower case letter.

Homozygous - (purebred) having two of the same genes present; an organism may be homozygous dominant with two dominant genes or homozygous recessive with two recessive genes.

Heterozygous - (hybrid) having one dominant gene and one recessive gene. The dominant gene will be expressed due to the Law of Dominance.

Genotype - the genes the organism has. Genes are represented with letters. AA, Bb, and tt are examples of genotypes.

Phenotype - how the trait is expressed in an organism. Blue eyes, brown hair, and red flowers are examples of phenotypes.

Incomplete dominance - neither gene masks the other; a new phenotype is formed. For example, red flowers and white flowers may have equal strength. A heterozygote (Rr) would have pink flowers. If a problem occurs with a third phenotype, incomplete dominance is occurring.

Codominance - genes may form new phenotypes. The ABO blood grouping is an example of co-dominance. A and B are of equal strength and O is recessive. Therefore, type A blood may have the genotypes of AA or AO, type B blood may have the genotypes of BB or BO, type AB blood has the genotype A and B, and type O blood has two recessive O genes.

Linkage - genes that are found on the same chromosome usually appear together unless crossing over has occurred in meiosis. (Example - blue eyes and blonde hair)

Lethal alleles - these are usually recessive due to the early death of the offspring. If a 2:1 ratio of alleles is found in offspring, a lethal gene combination is usually the reason. Some examples of lethal alleles include sickle cell anemia, Tay-Sachs and cystic fibrosis. Usually the coding for an important protein is affected.

Inborn errors of metabolism - these occur when the protein affected is an enzyme. Examples include PKU (phenylketonuria) and albanism.

Polygenic characters - many alleles code for a phenotype. There may be as many as twenty genes that code for skin color. This is why there is such a variety of skin tones. Another example is height. A couple of medium height may have very tall offspring.

Sex linked traits - the Y chromosome found only in males (XY) carries very little genetic information, whereas the X chromosome found in females (XX) carries very important information. Since men have no second X chromosome to cover up a recessive gene, the recessive trait is expressed more often in men. Women need the recessive gene on both X chromosomes to show the trait. Examples of sex linked traits include hemophilia and color-blindness.

Sex influenced traits - traits are influenced by the sex hormones. Male pattern baldness is an example of a sex influenced trait. Testosterone influences the expression of the gene. Usually men loose their hair due to this trait.

Competency 0035 Understand principles and theories related to biological evolution

Darwin defined the theory of Natural Selection in the mid-1800's. Through the study of finches on the Galapagos Islands, Darwin theorized that nature selects the traits that are advantageous to the organism. Those that do not possess the desirable trait die and do not pass on their genes. Those more fit to survive reproduce, thus increasing that gene in the population. Darwin listed four principles to define natural selection:

1. The individuals in a certain species vary from generation to generation.
2. Some of the variations are determined by the genetic makeup of the species.
3. More individuals are produced than will survive.
4. Some genes allow for better survival of an animal.

Causes of evolution - Certain factors increase the chances of variability in a population, thus leading to evolution. Items that increase variability include mutations, sexual reproduction, immigration, and large population. Items that decrease variation would be natural selection, emigration, small population, and random mating.

Sexual selection - Genes that happen to come together determine the makeup of the gene pool. Animals that use mating behaviors may be successful or unsuccessful. An animal that lacks attractive plumage or has a weak mating call will not attract the female, thereby eventually limiting that gene in the gene pool. Mechanical isolation, where sex organs do not fit the female, has an obvious disadvantage.

Competency 0036 Understand characteristics of populations, communities, ecosystems, and biomes

Ecology is the study of organisms, where they live and their interactions with the environment. A **population** is a group of the same species in a specific area. A **community** is a group of populations residing in the same area. Communities that are ecologically similar with regard to temperature, rainfall and the species that live there are called **biomes**. Specific biomes include:

Marine - covers 75% of the earth. This biome is organized by the depth of the water. The intertidal zone is from the tide line to the edge of the water. The littoral zone is from the water's edge to the open sea. It includes coral reef habitats and is the most densely populated area of the marine biome. The open sea zone is divided into the epipelagic zone and the pelagic zone. The epipelagic zone receives more sunlight and has a larger number of species. The ocean floor is called the benthic zone and is populated with bottom feeders.

Tropical Rain Forest - temperature is constant (25 degrees C), rainfall exceeds 200 cm. per year. Located around the area of the equator, the rain forest has abundant, diverse species of plants and animals.

Savanna - temperatures range from 0-25 degrees C depending on the location. Rainfall is from 90 to 150 cm per year. Plants include shrubs and grasses. The savanna is a transitional biome between the rain forest and the desert.

Desert - temperatures range from 10-38 degrees C. Rainfall is under 25 cm per year. Plant species include xerophytes and succulents. Lizards, snakes and small mammals are common animals.

Temperate Deciduous Forest - temperature ranges from -24 to 38 degrees C. Rainfall is between 65 to 150 cm per year. Deciduous trees are common, as well as deer, bear and squirrels.

Taiga - temperatures range from -24 to 22 degrees C. Rainfall is between 35 to 40 cm per year. Taiga is located very north and very south of the equator, getting close to the poles. Plant life includes conifers and plants that can withstand harsh winters. Animals include weasels, mink, and moose.

Tundra - temperatures range from -28 to 15 degrees C. Rainfall is limited, ranging from 10 to 15 cm per year. The tundra is located even further north and south than the taiga. Common plants include lichens and mosses. Animals include polar bears and musk ox.

Polar or Permafrost - temperature ranges from -40 to 0 degrees C. It rarely gets above freezing. Rainfall is below 10 cm per year. Most water is bound up as ice. Life is limited.

Succession - Succession is an orderly process of replacing a community that has been damaged or beginning one where no life previously existed. Primary succession occurs after a community has been totally wiped out by a natural disaster or where life never existed before, as in a flooded area. Secondary succession takes place in communities that were once flourishing but were disturbed by some source, either man or nature, but were not totally stripped. A climax community is a community that is established and flourishing.

Feeding relationships:

Parasitism - two species that occupy a similar place; the parasite benefits from the relationship, the host is harmed.

Commensalism - two species that occupy a similar place; neither species is harmed or benefits from the relationship.

Mutualism (symbiosis) - two species that occupy a similar place; both species benefit from the relationship.

Competition - two species that occupy the same habitat or eat the same food are said to be in competition with each other.

Predation - animals that eat other animals are called predators. The animals they feed on are called the prey. Population growth depends upon competition for food, water, shelter, and space. The amount of predators determines the amount of prey which, in turn, affects the number of predators.

Carrying Capacity - this is the total amount of life a habitat can support. Once the habitat runs out of food, water, shelter, or space, the carrying capacity decreases and then stabilizes.

Ecological Problems - nonrenewable resources are fragile and must be conserved for use in the future. Man's impact and knowledge of conservation will control our future.

Biological magnification - chemicals and pesticides accumulate along the food chain. Tertiary consumers have more accumulated toxins than animals at the bottom of the food chain.

Simplification of the food web - Three major crops feed the world (rice, corn, wheat). The planting of these foods wipe out habitats and push animals residing there into other habitats causing overpopulation or extinction.

Fuel sources - strip mining and the overuse of oil reserves have depleted these resources. At the current rate of consumption, conservation or alternate fuel sources will guarantee our future fuel sources.

Pollution - although technology gives us many advances, pollution is a side effect of production. Waste disposal and the burning of fossil fuels have polluted our land, water and air. Global warming and acid rain are two results of the burning of hydrocarbons and sulfur.

Global warming - rainforest depletion and the use of fossil fuels and aerosols have caused an increase in carbon dioxide production. This leads to a decrease in the amount of oxygen which is directly proportional to the amount of ozone. As the ozone layer depletes, more heat enters our atmosphere and is trapped. This causes an overall warming effect which may eventually melt polar ice caps, causing a rise in water levels and changes in climate which will affect weather systems world-wide.

Endangered species - construction of homes to house people in our overpopulated world has caused the destruction of habitat for other animals, leading to their extinction.

Overpopulation - the human race is still growing at an exponential rate. Carrying capacity has not been met due to our ability to use technology to produce more food and housing. Space and water can not be manufactured and eventually our non-renewable resources will reach a crisis state. Our overuse affects every living thing on this planet.

Biotic factors - living things in an ecosystem; plants, animals, bacteria, fungi, etc. If one population in a community increases, it affects the ability of another population to succeed by limiting the available amount of food, water, shelter and space.

Abiotic factors - non-living aspects of an ecosystem; soil quality, rainfall, and temperature. Changes in climate and soil can cause effects at the beginning of the food chain, thus limiting or accelerating the growth of populations.

SUBAREA X. EARTH AND SPACE SCIENCE

Competency 0037 Understand geological history and processes related to the changing earth

The biological history of the earth is partitioned into four major eras that are further divided into major periods. The latter periods are refined into groupings called Epochs.

Earth's history extends over more than four billion years and is reckoned in terms of a scale. Paleontologists who study the history of the Earth have divided this huge period of time into four large time units called eons. Eons are divided into smaller units of time called eras. An era refers to a time interval in which particular plants and animals were dominant or present in great abundance. The end of an era is most often characterized by (1) a general uplifting of the crust, (2) the extinction of the dominant plants or animals, and (3) the appearance of new life forms.

Each era is divided into several smaller divisions of time called periods. Some periods are divided into smaller time units called epochs.

Methods of geologic dating

Estimates of the Earth's age have been made possible with the discovery of **radioactivity** and the invention of instruments that can measure the amount of radioactivity in rocks. The use of radioactivity to make accurate determinations of Earth's age is called absolute dating. This process depends upon comparing the amount of radioactive material in a rock with the amount that has decayed into another element. Studying the radiation given off by atoms of radioactive elements is the most accurate method of measuring the Earth's age. These atoms are unstable and are continuously breaking down or undergoing decay. The radioactive element that decays is called the parent element. The new element that results from the radioactive decay of the parent element is called the daughter element.

The time required for one half of a given amount of a radioactive element to decay is called the half-life of that element or compound.

Geologists commonly use carbon dating to calculate the age of a fossil substance.

Infer the history of an area using geologic evidence

The determination of the age of rocks by cataloging their composition has been outmoded since the middle 1800s. Today, a sequential history can be determined by the fossil content (principle of fossil succession) of a rock system as well as its superposition within a range of systems. This classification process was termed stratigraphy and permitted the construction of a geologic column in which rock systems are arranged in their correct chronological order.

Principles of catastrophism and uniformitarianism

Uniformitarianism - is a fundamental concept in modern geology. It simply states that the physical, chemical, and biological laws that operated in the geologic past operate in the same way today. The forces and processes that we observe presently shaping our planet have been at work for a very long time. This idea is commonly stated as "the present is the key to the past."

Catastrophism - the concept that the earth was shaped by catastrophic events of a short term nature.

Orogeny is the term given to natural mountain building.

A mountain is terrain that has been raised high above the surrounding landscape by volcanic action, or some form of tectonic plate collisions. The plate collisions could be intercontinental or ocean floor collisions with a continental crust (subduction). The physical composition of mountains would include igneous, metamorphic, or sedimentary rocks; some may have rock layers that are tilted or distorted by plate collision forces.

There are many different types of mountains. The physical attributes of a mountain range depends upon the angle at which plate movement thrusts layers of rock to the surface. Many mountains (Adirondacks, Southern Rockies) were formed along high angle faults.

Folded mountains (Alps, Himalayas) are produced by the folding of rock layers during their formation. The Himalayas are the highest mountains in the world and contain Mount Everest which rises almost 9 km above sea level. The Himalayas were formed when India collided with Asia. The movement which created this collision is still in process at the rate of a few centimeters per year.

Fault-block mountains (Utah, Arizona, and New Mexico) are created when plate movement produces tension forces instead of compression forces. The area under tension produces normal faults and rock along these faults is displaced upward.

Dome mountains are formed as magma tries to push up through the crust but fails to break the surface. Dome mountains resemble a huge blister on the earth's surface.

Upwarped mountains (Black Hills of S.D.) are created in association with a broad arching of the crust. They can also be formed by rock thrusting upward along high angle faults.

Volcanism is the term given to the movement of magma through the crust and its emergence as lava onto the earth's surface. Volcanic mountains are built up by successive deposits of volcanic materials.

An active volcano is one that is presently erupting or building to an eruption. A dormant volcano is one that is between eruptions but still shows signs of internal activity that might lead to an eruption in the future. An extinct volcano is said to be no longer capable of erupting. Most of the world's active volcanoes are found along the rim of the Pacific Ocean, which is also a major earthquake zone. This curving belt of active faults and volcanoes is often called the Ring of Fire.

The world's best known volcanic mountains include: Mount Etna in Italy and Mount Kilimanjaro in Africa. The Hawaiian Islands are actually the tops of a chain of volcanic mountains that rise from the ocean floor.

There are three types of volcanic mountains: shield volcanoes, cinder cones and composite volcanoes.

Shield Volcanoes are associated with quiet eruptions. Lava emerges from the vent or opening in the crater and flows freely out over the earth's surface until it cools and hardens into a layer of igneous rock. A repeated lava flow builds this type of volcano into the largest volcanic mountain. Mauna Loa, found in Hawaii, is the largest volcano on earth.

Cinder Cone Volcanoes are associated with explosive eruptions as lava is hurled high into the air in a spray of droplets of various sizes. These droplets cool and harden into cinders and particles of ash before falling to the ground. The ash and cinder pile up around the vent to form a steep, cone-shaped hill called the cinder cone. Cinder cone volcanoes are relatively small but may form quite rapidly.

Composite Volcanoes are described as being built by both lava flows and layers of ash and cinders. Mount Fuji in Japan, Mount St. Helens in Washington, USA and Mount Vesuvius in Italy are all famous composite volcanoes.

Mechanisms of producing mountains

Mountains are produced by different types of mountain-building processes. Most major mountain ranges are formed by the processes of folding and faulting.

Folded Mountains are produced by the folding of rock layers. Crustal movements may press horizontal layers of sedimentary rock together from the sides, squeezing them into wavelike folds. Up-folded sections of rock are called anticlines; down-folded sections of rock are called synclines. The Appalachian Mountains are an example of folded mountains with long ridges and valleys in a series of anticlines and synclines formed by folded rock layers.

Faults are fractures in the earth's crust which have been created by either tension or compressive forces transmitted through the crust. These forces are produced by the movement of separate blocks of crust.

Faultings are categorized on the basis of the relative movement between the blocks on both sides of the fault plane. The movement can be horizontal, vertical or oblique.

A dip-slip fault occurs when the movement of the plates is vertical and opposite. The displacement is in the direction of the inclination, or dip, of the fault. Dip-slip faults are classified as normal faults when the rock above the fault plane moves down relative to the rock below.

Reverse faults are created when the rock above the fault plane moves up relative to the rock below. Reverse faults having a very low angle to the horizontal are also referred to as thrust faults.

Faults in which the dominant displacement is horizontal movement along the trend or strike (length) of the fault are called **strike-slip faults**. When a large strike-slip fault is associated with plate boundaries it is called a **transform fault**. The San Andreas Fault in California is a well-known transform fault.

Faults that have both vertical and horizontal movement are called **oblique-slip faults**.

When lava cools, igneous rock is formed. This formation can occur either above ground or below ground.

Intrusive rock includes any igneous rock that was formed below the earth's surface. Batholiths are the largest structures of intrusive type rock and are composed of near granite materials; they are the core of the Sierra Nevada Mountains.

Extrusive rock includes any igneous rock that was formed at the earth's surface.

Dikes are old lava tubes formed when magma entered a vertical fracture and hardened. Sometimes magma squeezes between two rock layers and hardens into a thin horizontal sheet called a **sill**. A **laccolith** is formed in much the same way as a sill, but the magma that creates a laccolith is very thick and does not flow easily. It pools and forces the overlying strata creating an obvious surface dome.

A **caldera** is normally formed by the collapse of the top of a volcano. This collapse can be caused by a massive explosion that destroys the cone and empties most if not all of the magma chamber below the volcano. The cone collapses into the empty magma chamber forming a caldera.

An inactive volcano may have magma solidified in its pipe. This structure, called a volcanic neck, is resistant to erosion and today may be the only visible evidence of the past presence of an active volcano.

When lava cools, igneous rock is formed. This formation can occur either above ground or below ground.

Glaciation

A continental glacier covered a large part of North America during the most recent ice age. Evidence of this glacial coverage remains as abrasive grooves, large boulders from northern environments dropped in southerly locations, glacial troughs created by the rounding out of steep valleys by glacial scouring, and the remains of glacial sources called **cirques** that were created by frost wedging the rock at the bottom of the glacier. Remains of plants and animals found in warm climate have been discovered in the moraines and outwash plains help to support the theory of periods of warmth during the past ice ages.

The Ice Age began about 2 -3 million years ago. This age saw the advancement and retreat of glacial ice over millions of years. Theories relating to the origin of glacial activity include Plate Tectonics, where it can be demonstrated that some continental masses, now in temperate climates, were at one time blanketed by ice and snow. Another theory involves changes in the earth's orbit around the sun, changes in the angle of the earth's axis, and the wobbling of the earth's axis. Support for the validity of this theory has come from deep ocean research that indicates a correlation between climatic sensitive micro-organisms and the changes in the earth's orbital status.

About 12,000 years ago, a vast sheet of ice covered a large part of the northern United States. This huge, frozen mass had moved southward from the northern regions of Canada as several large bodies of slow-moving ice, or glaciers. A time period in which glaciers advance over a large portion of a continent is called an ice age. A glacier is a large mass of ice that moves or flows over the land in response to gravity. Glaciers form among high mountains and in other cold regions.

There are two main types of glaciers: valley glaciers and continental glaciers. Erosion by valley glaciers is characteristic of U-shaped erosion. They produce sharp peaked mountains such as the Matterhorn in Switzerland. Erosion by continental glaciers often rides over mountains in their paths leaving smoothed, rounded mountains and ridges.

Plate tectonics

Data obtained from many sources led scientists to develop the theory of plate tectonics. This theory is the most current model that explains not only the movement of the continents, but also the changes in the earth's crust caused by internal forces.

Plates are rigid blocks of earth's crust and upper mantle. These rigid solid blocks make up the lithosphere. The earth's lithosphere is broken into nine large sections and several small ones. These moving slabs are called plates. The major plates are named after the continents they are "transporting".

The plates float on and move with a layer of hot, plastic-like rock in the upper mantle. Geologists believe that the heat currents circulating within the mantle cause this plastic zone of rock to slowly flow, carrying along the overlying crustal plates.

Movement of these crustal plates creates areas where the plates diverge as well as areas where the plates converge. A major area of divergence is located in the Mid-Atlantic. Currents of hot mantle rock rise and separate at this point of divergence creating new oceanic crust at the rate of 2 to 10 centimeters per year. Convergence is when the oceanic crust collides with either another oceanic plate or a continental plate. The oceanic crust sinks, forming an enormous trench and generating volcanic activity. Convergence also includes continent-to-continent plate collisions. When two plates slide past one another a transform fault is created.

These movements produce many major features of the earth's surface, such as mountain ranges, volcanoes, and earthquake zones. Most of these features are located at plate boundaries, where the plates interact by spreading apart, pressing together, or sliding past each other. These movements are very slow, averaging only a few centimeters a year.

Boundaries form between spreading plates where the crust is forced apart in a process called rifting. Rifting generally occurs at mid-ocean ridges. Rifting can also take place within a continent, splitting the continent into smaller landmasses that drift away from each other, thereby forming an ocean basin between them. The Red Sea is a product of rifting. As the seafloor spreading takes place, new material is added to the inner edges of the separating plates. In this way the plates grow larger, and the ocean basin widens. This is the process that broke up the super-continent Pangaea and created the Atlantic Ocean.

Boundaries between plates that are colliding are zones of intense crustal activity. When a plate of ocean crust collides with a plate of continental crust, the more dense oceanic plate slides under the lighter continental plate and plunges into the mantle. This process is called **subduction**, and the site where it takes place is called a subduction zone. A subduction zone is usually seen on the sea-floor as a deep depression called a trench.

The crustal movement, which is identified by plates sliding sideways past each other, produces a plate boundary characterized by major faults that are capable of unleashing powerful earth-quakes. The San Andreas Fault forms such a boundary between the Pacific Plate and the North American Plate.

Three major subdivisions of rocks are sedimentary, metamorphic and igneous.

Lithification of sedimentary rocks

When fluid sediments are transformed into solid sedimentary rocks, the process is known as **lithification**. One very common process affecting sediments is compaction where the weights of overlying materials compress and compact the deeper sediments. The compaction process leads to cementation. **Cementation** is when sediments are converted to sedimentary rock.

Factors in crystallization of igneous rocks

Igneous rocks can be classified according to their texture, their composition, and the way they were formed.

Molten rock is called magma. When molten rock pours out onto the surface of Earth, it is called lava.

As magma cools, the elements and compounds begin to form crystals. The slower the magma cools, the larger the crystals grow. Rocks with large crystals are said to have a coarse-grained texture. Granite is an example of a coarse grained rock. Rocks that cool rapidly before any crystals can form have a glassy texture such as obsidian, also commonly known as volcanic glass.

Metamorphic rocks are formed by high temperatures and great pressures. The process by which the rocks undergo these changes is called metamorphism. The outcome of metamorphic changes include deformation by extreme heat and pressure, compaction, destruction of the original characteristics of the parent rock, bending and folding while in a plastic stage, and the emergence of completely new and different minerals due to chemical reactions with heated water and dissolved minerals.

Metamorphic rocks are classified into two groups, foliated (leaf-like) rocks and unfoliated rocks. Foliated rocks consist of compressed, parallel bands of minerals, which give the rocks a striped appearance. Examples of such rocks include slate, schist, and gneiss. Unfoliated rocks are not banded and examples of such include quartzite, marble, and anthracite rocks.

Minerals are natural, non-living solids with a definite chemical composition and a crystalline structure. **Ores** are minerals or rock deposits that can be mined for a profit. **Rocks** are earth materials made of one or more minerals. A **Rock Facies** is a rock group that differs from comparable rocks (as in composition, age or fossil content).

Characteristics by which minerals are classified:

Minerals must adhere to five criteria. They must be (1) non-living, (2) formed in nature, (3) solid in form, (4) their atoms form a crystalline pattern, (5) its chemical composition is fixed within narrow limits.

There are over 3000 minerals in Earth's crust. Minerals are classified by composition. The major groups of minerals are silicates, carbonates, oxides, sulfides, sulfates, and halides. The largest group of minerals is the silicates. Silicates are made of silicon, oxygen, and one or more other elements.

Soils are composed of particles of sand, clay, various minerals, tiny living organisms, and humus, plus the decayed remains of plants and animals. Soils are divided into three classes according to their texture. These classes are sandy soils, clay soils, and loamy soils.

Sandy soils are gritty, and their particles do not bind together firmly. Sandy soils are porous- water passes through them rapidly. Sandy soils do not hold much water.

Clay soils are smooth and greasy, their particles bind together firmly. Clay soils are moist and usually do not allow water to pass through easily.

Loamy soils feel somewhat like velvet and their particles clump together. Loamy soils are made up of sand, clay, and silt. Loamy soils holds water but some water can pass through.

In addition to three main classes, soils are further grouped into three major types based upon their composition. These groups are pedalfers, pedocals, and laterites.

Pedalfers form in the humid, temperate climate of the eastern United States. Pedalfer soils contain large amounts of iron oxide and aluminum-rich clays, making the soil a brown to reddish brown color. This soil supports forest type vegetation.

Pedocals are found in the western United States where the climate is dry and temperate. These soils are rich in calcium carbonate. This type of soil supports grasslands and brush vegetation.

Laterites are found where the climate is wet and tropical. Large amounts of water flows through this soil. Laterites are red-orange soils rich in iron and aluminum oxides. There is little humus and this soil is not very fertile.

Erosion is the inclusion and transportation of surface materials by another moveable material, usually water, wind, or ice. The most important cause of erosion is running water. Streams, rivers, and tides are constantly at work removing weathered fragments of bedrock and carrying them away from their original location.

A stream erodes bedrock by the grinding action of the sand, pebbles and other rock fragments. This grinding against each other is called abrasion.

Streams also erode rocks by dissolving or absorbing their minerals. Limestone and marble are readily dissolved by streams.

The breaking down of rocks at or near to the earth's surface is known as **weathering**. Weathering breaks down these rocks into smaller and smaller pieces. There are two types of weathering: physical weathering and chemical weathering.

Physical weathering is the process by which rocks are broken down into smaller fragments without undergoing any change in chemical composition. Physical weathering is mainly caused by the freezing of water, the expansion of rock, and the activities of plants and animals.

Frost wedging is the cycle of daytime thawing and refreezing at night. This cycle causes large rock masses, especially the rocks exposed on mountain tops, to be broken into smaller pieces.

The peeling away of the outer layers from a rock is called exfoliation. Rounded mountain tops are called exfoliation domes and have been formed in this way.

Chemical weathering is the breaking down of rocks through changes in their chemical composition. An example would be the change of feldspar in granite to clay. Water, oxygen, and carbon dioxide are the main agents of chemical weathering. When water and carbon dioxide combine chemically, they produce a weak acid that breaks down rocks.

Decode map symbols

A system of imaginary lines has been developed that helps people describe exact locations on Earth. Looking at a globe of Earth, you will see lines drawn on it. The equator is drawn around Earth halfway between the North and South Poles. Latitude is a term used to describe distance in degrees north or south of the equator. Lines of latitude are drawn east and west parallel to the equator. Degrees of latitude range from 0 at the equator to 90 at either the North Pole or South Pole. Lines of latitude are also called parallels.

Lines drawn north and south at right angles to the equator and from pole to pole are called meridians. Longitude is a term used to describe distances in degrees east or west of a 0° meridian. The prime meridian is the 0° meridian and it passes through Greenwich, England.

Time zones are determined by longitudinal lines. Each time zone represents one hour. Since there are 24 hours in one complete rotation of the Earth, there are 24 international time zones. Each time zone is roughly 15° wide. While time zones are based on meridians, they do not strictly follow lines of longitude. Time zone boundaries are subject to political decisions and have been moved around cities and other areas at the whim of the electorate.

The International Date Line is the 180° meridian and it is on the opposite side of the world from the prime meridian. The International Date Line is one-half of one day or 12 time zones from the prime meridian. If you were traveling west across the International Date Line, you would lose one day. If you were traveling east across the International Date Line, you would gain one day.

Principles of contouring

A contour line is a line on a map representing an imaginary line on the ground that has the same elevation above sea level along its entire length. Contour intervals usually are given in even numbers or as a multiple of five. In mapping mountains, a large contour interval is used. Small contour intervals may be used where there are small differences in elevation.

Relief describes how much variation in elevation an area has. Rugged or high relief describes an area of many hills and valleys. Gentle or low relief describes a plains area or a coastal region. Five general rules should be remembered in studying contour lines on a map.

1. Contour lines close around hills and basins or depressions. Hachure lines are used to show depressions. Hachures are short lines placed at right angles to the contour line and they always point toward the lower elevation. A contour line that has hachures is called a depression contour.

2. Contours lines never cross. Contour lines are sometimes very close together. Each contour line represents a certain height above sea level.

3. Contour lines appear on both sides of an area where the slope reverses direction. Contour lines show where an imaginary horizontal plane would slice through a hillside or cut both sides of a valley.

4. Contours lines form V's that point upstream when they cross streams. Streams cut beneath the general elevation of the land surface, and contour lines follow a valley.

5. All contours lines either close (connect) or extend to the edge of the map. No map is large enough to have all its contour lines close.

Interpret maps and imagery

Like photographs, maps readily display information that would be impractical to express in words. Maps that show the shape of the land are called topographic maps. Topographic maps, which are also referred to as quadrangles, are generally classified according to publication scale. Relief refers to the difference in elevation between any two points. Maximum relief refers to the difference in elevation between the high and low points in the area being considered. Relief determines the contour interval, which is the difference in elevation between succeeding contour lines that are used on topographic maps.

Map scales express the relationship between distance or area on the map to the true distance or area on the earth's surface. It is expressed as so many feet (miles, meters, km, or degrees) per inch (cm) of map.

Competency 0038 Understand characteristics and properties of the hydrosphere

Seventy percent of the earth's surface is covered with saltwater which is termed the hydrosphere. The mass of this saltwater is about 1.4 x 10^{24} grams. The ocean waters continuously circulate among different parts of the hydrosphere. There are seven major oceans: the North Atlantic Ocean, South Atlantic Ocean, North Pacific Ocean, South Pacific Ocean, Indian Ocean, Arctic Ocean, and the Antarctic Ocean.

Pure water is a combination of the elements hydrogen and oxygen. These two elements make up about 96.5% of ocean water. The remaining portion is made up of dissolved solids. The concentration of these dissolved solids determines the water's salinity.

Salinity is the number of grams of these dissolved salts in 1,000 grams of sea water. The average salinity of ocean water is about 3.5%. In other words, one kilogram of sea water contains about 35 grams of salt. Sodium Chloride or salt (NaCl) is the most abundant of the dissolved salts. The dissolved salts also include smaller quantities of magnesium chloride, magnesium and calcium sulfates, and traces of several other salt elements. Salinity varies throughout the world oceans; the total salinity of the oceans varies from place to place and also varies with depth. Salinity is low near river mouths where the ocean mixes with fresh water, and salinity is high in areas of high evaporation rates.

The temperature of the ocean water varies with different latitudes and with ocean depths. Ocean water temperature is about constant to depths of 90 meters (m). The temperature of surface water will drop rapidly from 28° C at the equator to -2° C at the Poles. The freezing point of sea water is lower than the freezing point of pure water. Pure water freezes at 0° C. The dissolved salts in the sea water keep sea water at a freezing point of -2° C. The freezing point of sea water may vary depending on its salinity in a particular location.

The ocean can be divided into three temperature zones. The surface layer consists of relatively warm water and exhibits most of the wave action present. The area where the wind and waves churn and mix the water is called the mixed layer. This is the layer where most living creatures are found due to abundant sunlight and warmth. The second layer is called the thermocline and it becomes increasingly cold as its depth increases. This change is due to the lack of energy from sunlight. The layer below the thermocline continues to the deep dark, very cold, and semi-barren ocean floor.

Oozes - the name given to the sediment that contains at least 30% plant or animal shell fragments. Ooze contains calcium carbonate. Deposits that form directly from sea water in the place where they are found are called authigenic deposits. Maganese nodules are authigenic deposits found over large areas of the ocean floor.

Causes for the formation of ocean floor features.

The surface of the earth is in constant motion. This motion is the subject of Plate Tectonics studies. Major plate separation lines lie along the ocean floors. As these plates separate, molten rock rises, continuously forming new ocean crust and creating new and taller mountain ridges under the ocean. The Mid-Atlantic Range, which divides the Atlantic Ocean basin into two nearly equal parts, shows evidence from mapping of these deep-ocean floor changes.

Seamounts are formed by underwater volcanoes. Seamounts and volcanic islands are found in long chains on the ocean floor. They are formed when the movement of an oceanic plate positions a plate section over a stationary hot spot located deep in the mantle. Magma rising from the hot spot punches through the plate and forms a volcano. The Hawaiian Islands are examples of volcanic island chains.

Magma that rises to produce a curving chain of volcanic islands is called an island arc. An example of an island arc is the Lesser Antilles chain in the Caribbean Sea.

World weather patterns are greatly influenced by ocean surface currents in the upper layer of the ocean. These currents continuously move along the ocean surface in specific directions. Ocean currents that flow deep below the surface are called sub-surface currents. These currents are influenced by such factors as the location of landmasses in the current's path and the earth's rotation.

Surface currents are caused by winds and are classified by temperature. Cold currents originate in the Polar regions and flow through surrounding water that is measurably warmer. Those currents with a higher temperature than the surrounding water are called warm currents and can be found near the equator. These currents follow swirling routes around the ocean basins and the equator. The Gulf Stream and the California Current are the two main surface currents that flow along the coastlines of the United States. The Gulf Stream is a warm current in the Atlantic Ocean that carries warm water from the equator to the northern parts of the Atlantic Ocean. Benjamin Franklin studied and named the Gulf Stream. The California Current is a cold current that originates in the Arctic regions and flows southward along the west coast of the United States.

Differences in water density also create ocean currents. Water found near the bottom of oceans is the coldest and the densest. Water tends to flow from a denser area to a less dense area. Currents that flow because of a difference in the density of the ocean water are called density currents. Water with a higher salinity is denser than water with a lower salinity. Water that has salinity different from the surrounding water may form a density current.

Knowledge of the causes and effects of waves

The movement of ocean water is caused by the wind, the sun's heat energy, the earth's rotation, the moon's gravitational pull on earth, and by underwater earthquakes. Most ocean waves are caused by the impact of winds. Wind blowing over the surface of the ocean transfers energy (friction) to the water and causes waves to form. Waves are also formed by seismic activity on the ocean floor. A wave formed by an earthquake is called a seismic sea wave. These powerful waves can be very destructive, with wave heights increasing to 30 m or more near the shore. The crest of a wave is its highest point. The trough of a wave is its lowest point. The distance from wave top to wave top is the wavelength. The wave period is the time between the passings of two successive waves.

Seafloor

The ocean floor has many of the same features that are found on land. The ocean floor has higher mountains than are present on land, extensive plains and deeper canyons than are present on land. Oceanographers have named different parts of the ocean floor according to their structure. The major parts of the ocean floor are:

The **continental shelf** is the sloping part of the continent that is covered with water extending from the shoreline to the continental slope.

The **continental slope** is the steeply sloping area that connects the continental shelf and the deep-ocean floor.

The **continental rise** is the gently sloping surface at the base of the continental slope.

The **abyssal plains** are the flat, level parts of the ocean floor.

A **seamount** is an undersea volcano peak that is at least 1000 m above the ocean floor.

Guyot - A submerged flat-topped seamount

Mid-ocean ridges are continuous undersea mountain chains that are found mostly in the middle portions of the oceans.

Ocean trenches are long, elongated narrow troughs or depressions formed where ocean floors collide with another section of ocean floor or continent. The deepest trench in the Pacific Ocean is the Marianas Trench which is about 11 km deep.

Shoreline

The shoreline is the boundary where land and sea meet. Shorelines mark the average position of sea level, which is the average height of the sea without consideration of tides and waves. Shorelines are classified according to the way they were formed. The three types of shorelines are: submerged, emergent, and neutral. When the sea has risen, or the land has sunk, a **submerged shoreline** is created. An **emergent shoreline** occurs when sea falls or the land rises. A **neutral shoreline** does not show the features of a submerged or an emergent shoreline. A neutral shoreline is usually observed as a flat and broad beach. A **stack** is an island of resistant rock left after weaker rock is worn away by waves and currents. Waves approaching the beach at a slight angle create a current of water that flows parallel to the shore. This longshore current carries loose sediment almost like a river of sand. A spit is formed when a weak longshore current drops its load of sand as it turns into a bay.

Rip currents are narrow currents that flow seaward at a right angle to the shoreline. These currents are very dangerous to swimmers. Most of the beach sands are composed of grains of resistant material like quartz and orthoclase but coral or basalt are found in some locations. Many beaches have rock fragments that are too large to be classified as sand.

Water that falls to Earth in the form of rain and snow is called **precipitation.** Precipitation is part of a continuous process in which water at the Earth's surface evaporates, condenses into clouds, and returns to Earth. This process is termed the **water cycle**. The water located below the surface is called groundwater.

The impacts of altitude upon climatic conditions are primarily related to temperature and precipitation. As altitude increases, climatic conditions become increasingly drier and colder. Solar radiation becomes more severe as altitude increases while the effects of convection forces are minimized. Climatic changes as a function of latitude follow a similar pattern (as a reference, latitude moves either north or south from the equator). The climate becomes colder and drier as the distance from the equator increases. Proximity to land or water masses produces climatic conditions based upon the available moisture. Dry and arid climates prevail where moisture is scarce; lush tropical climates can prevail where moisture is abundant. Climate, as described above, depends upon the specific combination of conditions making up an area's environment. Man impacts all environments by producing pollutants in earth, air, and water. It follows then, that man is a major player in world climatic conditions.

Competency 0039 Understand the earth's atmosphere, weather, and climate

Air masses moving toward or away from the Earth's surface are called air currents. Air moving parallel to Earth's surface is called **wind**. Weather conditions are generated by winds and air currents carrying large amounts of heat and moisture from one part of the atmosphere to another. Wind speeds are measured by instruments called anemometers.

The wind belts in each hemisphere consist of convection cells that encircle Earth like belts. There are three major wind belts on Earth: (1) trade winds (2) prevailing westerlies, and (3) polar easterlies. Wind belt formation depends on the differences in air pressures that develop in the doldrums, the horse latitudes, and the polar regions. The Doldrums surround the equator. Within this belt heated air usually rises straight up into Earth's atmosphere. The Horse latitudes are regions of high barometric pressure with calm and light winds and the Polar regions contain cold dense air that sinks to the Earth's surface.

Winds caused by local temperature changes include sea breezes, and land breezes.

Sea breezes are caused by the unequal heating of the land and an adjacent, large body of water. Land heats up faster than water. The movement of cool ocean air toward the land is called a sea breeze. Sea breezes usually begin blowing about mid-morning and end about sunset.

A breeze that blows from the land to the ocean or a large lake is called a **land breeze.**

Monsoons are huge wind systems that cover large geographic areas and that reverse direction seasonally. The monsoons of India and Asia are examples of these seasonal winds. They alternate wet and dry seasons. As denser cooler air over the ocean moves inland, a steady seasonal wind called a summer or wet monsoon is produced.

Cloud types

Cirrus clouds - White and feathery; high in the sky

Cumulus – thick, white, fluffy

Stratus – layers of clouds cover most of the sky

Nimbus – heavy, dark clouds that represent thunderstorm clouds

Variation on the clouds mentioned above.

Cumulo-nimbus

Strato-nimbus

The air temperature at which water vapor begins to condense is called the **dew point.**

Relative humidity is the actual amount of water vapor in a certain volume of air compared to the maximum amount of water vapor this air could hold at a given temperature.

Knowledge of types of storms

A **thunderstorm** is a brief, local storm produced by the rapid upward movement of warm, moist air within a cumulo-nimbus cloud. Thunderstorms always produce lightning and thunder, and are accompanied by strong wind gusts and heavy rain or hail.

A severe storm with swirling winds that may reach speeds of hundreds of km per hour is called a **tornado**. Such a storm is also referred to as a "twister". The sky is covered by large cumulo-nimbus clouds and violent thunderstorms; a funnel-shaped swirling cloud may extend downward from a cumulo-nimbus cloud and reach the ground. Tornadoes are storms that leave a narrow path of destruction on the ground.

A swirling, funnel-shaped cloud that **extends** downward and touches a body of water is called a **waterspout.**

Hurricanes are storms that develop when warm, moist air carried by trade winds rotate around a low-pressure "eye". A large, rotating, low-pressure system accompanied by heavy precipitation and strong winds is called a tropical cyclone (better known as a hurricane). In the Pacific region, a hurricane is called a typhoon.

Storms that occur only in the winter are known as blizzards or ice storms. A **blizzard** is a storm with strong winds, blowing snow and frigid temperatures. An **ice storm** consists of falling rain that freezes when it strikes the ground, covering everything with a layer of ice.

Competency 0040 Understand components of the solar system and universe and their interactions

Two main hypotheses of the origin of the solar system are: (1) **the tidal hypothesis** and (2) **the condensation hypothesis**.

The tidal hypothesis proposes that the solar system began with a near collision of the sun and a large star. Some astronomers believe that as these two stars passed each other, the great gravitational pull of the large star extracted hot gases out of the sun. The mass from the hot gases started to orbit the sun, which began to cool then condensing into the nine planets. (Few astronomers support this example).

The condensation hypothesis proposes that the solar system began with rotating clouds of dust and gas. Condensation occurred in the center forming the sun and the smaller parts of the cloud formed the nine planets. (This example is widely accepted by many astronomers).

Two main theories to explain the origins of the universe include: (1) **The Big Bang Theory** and (2) **The Steady-State Theory.**

The Big Bang Theory has been widely accepted by many astronomers. It states that the universe originated from a magnificent explosion spreading mass, matter and energy into space. The galaxies formed from this material as it cooled during the next half-billion years.

The Steady-State Theory is the least accepted theory. It states that the universe is continuously being renewed. Galaxies move outward and new galaxies replace the older galaxies. Astronomers have not found any evidence to prove this theory.

The future of the universe is hypothesized with the Oscillating Universe Hypothesis. It states that the universe will oscillate or expand and contract. Galaxies will move away from one another and will in time slow down and stop. Then a gradual moving toward each other will again activate the explosion or The Big Bang theory.

The **sun** is considered the nearest star to earth that produces solar energy. By the process of nuclear fusion, hydrogen gas is converted to helium gas. Energy flows out of the core to the surface, then radiation escapes into space.

Parts of the sun include: (1) **core:** the inner portion of the sun where fusion takes place, (2) **photosphere:** considered the surface of the sun which produces **sunspots** (cool, dark areas that can be seen on its surface), (3) **chromosphere:** hydrogen gas causes this portion to be red in color (also found here are solar flares or sudden brightness of the chromosphere) and solar prominences (gases that shoot outward from the chromosphere), and (4) **corona**, the transparent area of sun visible only during a total eclipse.

Solar radiation is energy traveling from the sun that radiates into space. **Solar flares** produce excited protons and electrons that shoot outward from the chromosphere at great speeds reaching earth. These particles disturb radio reception and also affect the magnetic field on earth.

Knowledge of telescope types

Galileo was the first person to use telescopes to observe the solar system. He invented the first refracting telescope. A **refracting telescope** uses lenses to bend light rays to focus the image.

Sir Isaac Newton invented the **reflecting telescope** using mirrors to gather light rays on a curved mirror which produces a small focused image.

The world's largest telescope is located in Mauna Kea, Hawaii. It uses multiple mirrors to gather light rays.

The **Hubble Space telescope** uses a **single-reflector mirror**. It provides an opportunity for astronomers to observe objects seven times farther away. Even those objects that are 50 times fainter can be viewed better than by any telescope on Earth. There are future plans to make repairs and install new mirrors and other equipment on the Hubble Space telescope.

Refracting and reflecting telescopes are considered **optical telescopes** since they gather visible light and focus it to produce images. A different type of telescope that collects invisible radio waves created by the sun and stars is called a **radio telescope.**

Radio telescopes consists of a reflector or dish with special receivers. The reflector collects radio waves that are created by the sun and stars. Using a radio telescope has many advantages. They can receive signals 24 hours a day, can operate in any kind of weather, dust particles or clouds do not interfere with its performance. The most impressive aspect of the radio telescope is its ability to detect objects from such great distances in space.

The world's largest radio telescope is located in Arecibo, Puerto Rico. It has a collecting dish antenna of more than 300 meters in diameter.

Use spectral analysis to identify or infer features of stars or star systems

The **spectroscope** is a device or an attachment for telescopes that is used to separate white light into a series of different colors by wave lengths. This series of colors of light is called a **spectrum**. A **spectrograph** can photograph a spectrum. Wavelengths of light have distinctive colors. The color red has the longest wavelength and violet has the shortest wavelength. Wavelengths are arranged to form an **electromagnetic spectrum**. They range from very long radio waves to very short gamma rays. Visible light covers a small portion of the electromagnetic spectrum. Spectroscopes observe the spectra, temperatures, pressures, and also the movement of stars. The movements of stars indicate if they are moving toward, or away, from earth.

If a star is moving towards earth, light waves compress and the wavelengths of light seem shorter. This will cause the entire spectrum to move towards the blue or violet end of the spectrum.

If a star is moving away from earth, light waves expand and the wavelengths of light seem longer. This will cause the entire spectrum to move towards the red end of the spectrum.

Knowledge of astronomical measurement

The three formulas astronomers use for calculating distances in space are: (1) the **AU or astronomical unit**, (2) **the LY or Light year,** and (3) **the parsec**. It is important to remember that these formulas are measured in distances not time.

The distance between the earth and the sun is about 150×10^{6} km. This distance is known as an astronomical unit or AU. This formula is used to measure distances within the solar system, it is not used to measure time.

The distance light travels in one year is a light year (9.5×10^{12} km). This formula is used to measure distances in space, it does not measure time. Large distances are measured in parsecs. One parsec equals 3.26 light-years.

There are approximately 63,000 AUs in one light year or,

$$9.5 \times 10^{12} \text{ km} / 150 \times 10^{6} \text{ km} = 6.3 \times 10^{4} \text{ AU}$$

There are eight established planets in our solar system; Mercury, Venus, Earth, Mars, Jupiter, Saturn, Uranus, and Neptune. Pluto was an established planet in our solar system, but as of the summer of 2006, its status is being reconsidered. The planets are divided into two groups based on distance from the sun. The inner planets include: Mercury, Venus, Earth, and Mars. The outer planets include: Jupiter, Saturn, Uranus, and Neptune.

Planets

Mercury -- the closest planet to the sun. Its surface has craters and rocks. The atmosphere is composed of hydrogen, helium and sodium. Mercury was named after the Roman messenger god.

Venus -- has a slow rotation when compared to Earth. Venus and Uranus rotate in opposite directions from the other planets. This opposite rotation is called retrograde rotation. The surface of Venus is not visible due to the extensive cloud cover. The atmosphere is composed mostly of carbon dioxide. Sulfuric acid droplets in the dense cloud cover give Venus a yellow appearance. Venus has a greater greenhouse effect than observed on Earth. The dense clouds combined with carbon dioxide trap heat. Venus was named after the Roman goddess of love.

Earth -- considered a water planet with 70% of its surface covered by water. Gravity holds the masses of water in place. The different temperatures observed on earth allow for the different states (solid, liquid, gas) of water to exist. The atmosphere is composed mainly of oxygen and nitrogen. Earth is the only planet that is known to support life.

Mars -- the surface of Mars contains numerous craters, active and extinct volcanoes, ridges, and valleys with extremely deep fractures. Iron oxide found in the dusty soil makes the surface seem rust-colored and the skies seem pink in color. The atmosphere is composed of carbon dioxide, nitrogen, argon, oxygen and water vapor. Mars has polar regions with ice caps composed of water. Mars has two satellites. Mars was named after the Roman war god.

Jupiter -- largest planet in the solar system. Jupiter has 16 moons. The atmosphere is composed of hydrogen, helium, methane and ammonia. There are white colored bands of clouds indicating rising gas and dark colored bands of clouds indicating descending gases. The gas movement is caused by heat resulting from the energy of Jupiter's core. Jupiter has a Great Red Spot that is thought to be a hurricane-type cloud. Jupiter has a strong magnetic field.

Saturn -- the second largest planet in the solar system. Saturn has rings of ice, rock, and dust particles circling it. Saturn's atmosphere is composed of hydrogen, helium, methane, and ammonia. Saturn has 20 plus satellites. Saturn was named after the Roman god of agriculture.

Uranus -- the second largest planet in the solar system with retrograde revolution. Uranus is a gaseous planet. It has 10 dark rings and 15 satellites. Its atmosphere is composed of hydrogen, helium, and methane. Uranus was named after the Greek god of the heavens.

Neptune -- another gaseous planet with an atmosphere consisting of hydrogen, helium, and methane. Neptune has 3 rings and 2 satellites. Neptune was named after the Roman sea god because its atmosphere is the same color as the seas.

Pluto – once considered the smallest planet in the solar system; its status as a planet is being reconsidered. Pluto's atmosphere probably contains methane, ammonia, and frozen water. Pluto has 1 satellite. Pluto revolves around the sun every 250 years. Pluto was named after the Roman god of the underworld.

Comets, asteroids, and meteors

Astronomers believe that rocky fragments may have been the remains of the birth of the solar system that never formed into a planet. **Asteroids** are found in the region between Mars and Jupiter.

Comets are masses of frozen gases, cosmic dust, and small rocky particles. Astronomers think that most comets originate in a dense comet cloud beyond Pluto. Comet consists of a nucleus, a coma, and a tail. A comet's tail always points away from the sun. The most famous comet, **Halley's Comet,** is named after the person whom first discovered it in 240 B.C. It returns to the skies near earth every 75 to 76 years.

Meteoroids are composed of particles of rock and metal of various sizes. When a meteoroid travels through the earth's atmosphere, friction causes its surface to heat up and it begins to burn. The burning meteoroid falling through the earth's atmosphere is called a **meteor** (also known as a "shooting star").

Meteorites are meteors that strike the earth's surface. A physical example of a meteorite's impact on the earth's surface can be seen in Arizona. The Barringer Crater is a huge meteor crater. There are many other meteor craters throughout the world.

Astronomers use groups or patterns of stars called **constellations** as reference points to locate other stars in the sky. Familiar constellations include: Ursa Major (also known as the big bear) and Ursa Minor (known as the little bear). Within the Ursa Major, the smaller constellation, The Big Dipper is found. Within the Ursa Minor, the smaller constellation, The Little Dipper is found.

Different constellations appear as the earth continues its revolution around the sun with the seasonal changes.

Magnitude stars are 21 of the brightest stars that can be seen from earth. These are the first stars noticed at night. In the Northern Hemisphere there are 15 commonly observed first magnitude stars.

A vast collection of stars are defined as **galaxies**. Galaxies are classified as irregular, elliptical, and spiral. An irregular galaxy has no real structured appearance; most are in their early stages of life. An elliptical galaxy consists of smooth ellipses, containing little dust and gas, but composed of millions or trillion stars. Spiral galaxies are disk-shaped and have extending arms that rotate around its dense center. Earth's galaxy is found in the Milky Way and it is a spiral galaxy.

Terms related to deep space

A **pulsar** is defined as a variable radio source that emits signals in very short, regular bursts; it is believed to be a rotating neutron star.

A **quasar** is defined as an object that photographs like a star but has an extremely large redshift and a variable energy output; believed to be the active core of a very distant galaxy.

Black holes are defined as an object that has collapsed to such a degree that light cannot escape from its surface. Light is trapped by the intense gravitational field.

SUBAREA XI. INTEGRATION OF KNOWLEDGE AND UNDERSTANDING

Competency 0041 Prepare an organized, developed analysis on a topic related to one or more of the following: number sense and operations; patterns, relations, and algebra; geometry and measurement; data analysis, statistics, and probability; and trigonometry, calculus, and discrete mathematics.

Exercise: Interpreting Slope as a Rate of Change
Connection: Social Sciences/Geography

Real-life Application: Slope is often used to describe a constant or average rate of change. These problems usually involve units of measure such as miles per hour or dollars per year.

Problem:

The town of Verdant Slopes has been experiencing a boom in population growth. By the year 2000, the population had grown to 45,000, and by 2005, the population had reached 60,000.

Communicating about Algebra:

a. Using the formula for slope as a model, find the average rate of change in population growth, expressing your answer in people per year.

Extension:

b. Using the average rate of change determined in a., predict the population of Verdant Slopes in the year 2010.

Solution:

a. Let t represent the time and p represent population growth. The two observances are represented by (t_1, p_1) and (t_2, p_2).

1st observance = (t_1, p_1) = (2000, 45000)
2nd observance = (t_2, p_2) = (2005, 60000)

Use the formula for slope to find the average rate of change.

$$\text{Rate of change} = \frac{p_2 - p_1}{t_2 - t_1}$$

Substitute values.

$$= \frac{60000 - 45000}{2005 - 2000}$$

Simplify.

$$= \frac{15000}{5} = 3000\, people/year$$

The average rate of change in population growth for Verdant Slopes between the years 2000 and 2005 was 3000 people/year.

b.

$$3000\, people/year \times 5\, years = 15000\, people$$
$$60000\, people + 15000\, people = 75000\, people$$

At a continuing average rate of growth of 3000 people/year, the population of Verdant Slopes could be expected to reach 75,000 by the year 2010.

SUBAREA XII. INTEGRATION OF KNOWLEDGE AND UNDERSTANDING OF SCIENCE

Competency 0042 Prepare an organized, developed analysis on a topic related to one or more of the following: history, philosophy, and methodology of science; chemistry; physics; biology; and earth and space science

Discuss the scientific process.

Best:

Science may be defined as a body of knowledge that is systematically derived from study, observations, and experimentation. Its goal is to identify and establish principles and theories that may be applied to solve problems. Pseudoscience, on the other hand, is a belief that is not warranted. There is no scientific methodology or application. Some of the more classic examples of pseudoscience include witchcraft, alien encounters or any topic that is explained by hearsay.

Scientific theory and experimentation must be repeatable. It is also possible to be disproved and is capable of change. Science depends on communication, agreement, and disagreement among scientists. It is composed of theories, laws, and hypotheses.

theory - the formation of principles or relationships which have been verified and accepted.

law - an explanation of events that occur with uniformity under the same conditions (laws of nature, law of gravitation).

hypothesis - an unproved theory or educated guess followed by research to best explain a phenomena. A theory is a proven hypothesis.

Science is limited by the available technology. An example of this would be the relationship of the discovery of the cell and the invention of the microscope. As our technology improves, more hypotheses will become theories and possibly laws. Science is also limited by the data that is able to be collected. Data may be interpreted differently on different occasions. Science limitations cause explanations to be changeable as new technologies emerge.

The first step in scientific inquiry is posing a question to be answered. Next, a hypothesis is formed to provide a plausible explanation. An experiment is then proposed and performed to test this hypothesis. A comparison between the predicted and observed results is the next step. Conclusions are then formed and it is determined whether the hypothesis is correct or incorrect. If incorrect, the next step is to form a new hypothesis and the process is repeated.

Better:

Science is derived from study, observations, and experimentation. Its goal is to identify and establish principles and theories that may be applied to solve problems. Scientific theory and experimentation must be repeatable. It is also possible to disprove or change a theory. Science depends on communication, agreement, and disagreement among scientists. It is composed of theories, laws, and hypotheses.

A theory is a principle or relationship that has been verified and accepted through experiments. A law is an explanation of events that occur with uniformity under the same conditions. A hypothesis is an educated guess followed by research. A theory is a proven hypothesis.

Science is limited by the available technology. An example of this would be the relationship of the discovery of the cell and the invention of the microscope. The first step in scientific inquiry is posing a question to be answered. Next, a hypothesis is formed to provide a plausible explanation. An experiment is then proposed and performed to test this hypothesis. A comparison between the predicted and observed results is the next step. Conclusions are then formed and it is determined whether the hypothesis is correct or incorrect. If incorrect, the next step is to form a new hypothesis and the process is repeated.

Basic:

Science is composed of theories, laws, and hypotheses. The first step in scientific inquiry is posing a question to be answered. Next, a hypothesis is formed to provide a plausible explanation. An experiment is then proposed and performed to test this hypothesis. A comparison between the predicted and observed results is the next step. Conclusions are then formed and it is determined whether the hypothesis is correct or incorrect. If incorrect, the next step is to form a new hypothesis and the process is repeated. Science is always limited by the available technology.

Sample Test: Science

DIRECTIONS: Read each item and select the correct response. The answer key follows.

1. In an experiment, the scientist states that he believes a change in the color of a liquid is due to a change of pH. This is an example of _________ .

A. observing.

B. inferring.

C. measuring.

D. classifying.

2. When is a hypothesis formed?

A. Before the data is taken.

B. After the data is taken.

C. After the data is analyzed.

D. Concurrent with graphing the data.

3. Who determines the laws regarding the use of safety glasses in the classroom?

A. The state.

B. The school site.

C. The Federal government.

D. The district level.

4. If one inch equals 2.54 cm how many mm in 1.5 feet? (APPROXIMATELY)

A. 18 mm.

B. 1800 mm.

C. 460 mm.

D. 4,600 mm.

5. Which of the following instruments measures wind speed?

A. A barometer.

B. An anemometer.

C. A wind sock.

D. A weather vane.

6. Sonar works by _________ .

A. timing how long it takes sound to reach a certain speed.

B. bouncing sound waves between two metal plates.

C. bouncing sound waves off an underwater object and timing how long it takes for the sound to return.

D. evaluating the motion and amplitude of sound.

7. The measure of the pull of the earth's gravity on an object is called _________.

A. mass number.

B. atomic number.

C. mass.

D. weight.

8. Which reaction below is a decomposition reaction?

A. $HCl + NaOH \rightarrow NaCl + H_2O$

B. $C + O_2 \rightarrow CO_2$

C. $2H_2O \rightarrow 2H_2 + O_2$

D. $CuSO_4 + Fe \rightarrow FeSO_4 + Cu$

9. The Law of Conservation of Energy states that _________.

A. There must be the same number of products and reactants in any chemical equation.

B. Objects always fall toward large masses such as planets.

C. Energy is neither created nor destroyed, but may change form.

D. Lights must be turned off when not in use, by state regulation.

10. Which parts of an atom are located inside the nucleus?

A. electrons and neutrons.

B. protons and neutrons.

C. protons only.

D. neutrons only.

11. The elements in the modern Periodic Table are arranged _________.

A. in numerical order by atomic number.

B. randomly.

C. in alphabetical order by chemical symbol.

D. in numerical order by atomic mass.

12. Carbon bonds with hydrogen by _________.

A. ionic bonding.

B. non-polar covalent bonding.

C. polar covalent bonding.

D. strong nuclear force.

13. Vinegar is an example of a __________ .

A. strong acid.

B. strong base.

C. weak acid.

D. weak base.

14. Which of the following is not a nucleotide?

A. adenine.

B. alanine.

C. cytosine.

D. guanine.

15. When measuring the volume of water in a graduated cylinder, where does one read the measurement?

A. At the highest point of the liquid.

B. At the bottom of the meniscus curve.

C. At the closest mark to the top of the liquid.

D. At the top of the plastic safety ring.

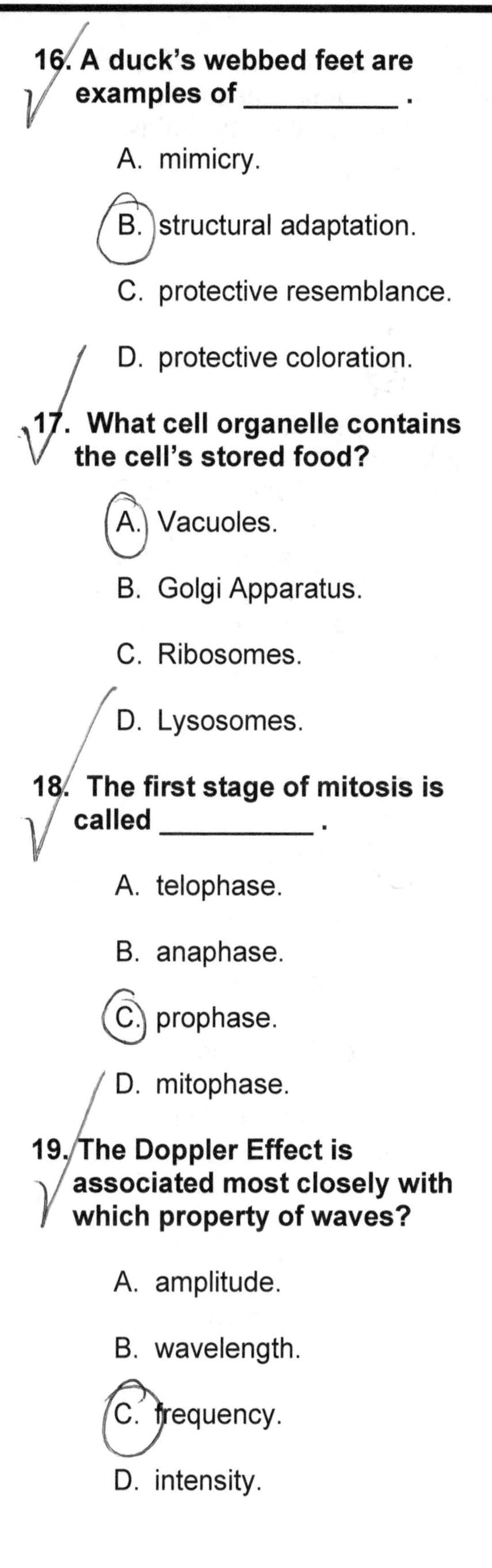

16. A duck's webbed feet are examples of __________ .

A. mimicry.

B. structural adaptation.

C. protective resemblance.

D. protective coloration.

17. What cell organelle contains the cell's stored food?

A. Vacuoles.

B. Golgi Apparatus.

C. Ribosomes.

D. Lysosomes.

18. The first stage of mitosis is called __________ .

A. telophase.

B. anaphase.

C. prophase.

D. mitophase.

19. The Doppler Effect is associated most closely with which property of waves?

A. amplitude.

B. wavelength.

C. frequency.

D. intensity.

20. Viruses are responsible for many human diseases including all of the following *except* __________ ?

A. influenza.

B. A.I.D.S.

C. the common cold.

D. strep throat.

21. A series of experiments on pea plants formed by _________ showed that two invisible markers existed for each trait, and one marker dominated the other.

A. Pasteur.

B. Watson and Crick.

C. Mendel.

D. Mendeleev.

22. Formaldehyde should not be used in school laboratories for the following reason:

A. it smells unpleasant.

B. it is a known carcinogen.

C. it is expensive to obtain.

D. it is explosive.

23. Amino acids are carried to the ribosome in protein synthesis by __________ .

A. transfer RNA (tRNA).

B. messenger RNA (mRNA).

C. ribosomal RNA (rRNA).

D. transformation RNA (trRNA).

24. When designing a scientific experiment, a student considers all the factors that may influence the results. The process goal is to ________.

A. recognize and manipulate independent variables.

B. recognize and record independent variables.

C. recognize and manipulate dependent variables.

D. recognize and record dependent variables.

Almost!!!! Thought it was this!

25. Since ancient times, people have been entranced with bird flight. What is the key to bird flight?

A. Bird wings are a particular shape and composition.

B. Birds flap their wings quickly enough to propel themselves.

C. Birds take advantage of tailwinds.

D. Birds take advantage of crosswinds.

26. Laboratory researchers have classified fungi as distinct from plants because the cell walls of fungi __________ .

A. contain chitin.

B. contain yeast.

C. are more solid.

D. are less solid.

27. In a fission reactor, "heavy water" is used to __________ .

A. terminate fission reactions.

B. slow down neutrons and moderate reactions.

C. rehydrate the chemicals.

D. initiate a chain reaction.

28. The transfer of heat by electromagnetic waves is called __________ .

A. conduction.

B. convection.

C. phase change.

D. radiation.

29. When heat is added to most solids, they expand. Why is this the case?

A. The molecules get bigger.

B. The faster molecular motion leads to greater distance between the molecules.

C. The molecules develop greater repelling electric forces.

D. The molecules form a more rigid structure.

30. The force of gravity on earth causes all bodies in free fall to __________ .

A. fall at the same speed.

B. accelerate at the same rate.

C. reach the same terminal velocity.

D. move in the same direction.

31. Sound waves are produced by __________ .

A. pitch.

B. noise.

C. vibrations.

D. sonar.

32. Resistance is measured in units called __________ .

A. watts.

B. volts.

C. ohms.

D. current.

33. Sound can be transmitted in all of the following *except* __________ .

A. air.

B. water.

C. A diamond.

D. a vacuum.

34. As a train approaches, the whistle sounds __________ .

A. higher, because it has a higher apparent frequency.

B. lower, because it has a lower apparent frequency.

C. higher, because it has a lower apparent frequency.

D. lower, because it has a higher apparent frequency.

35. The speed of light is different in different materials. This is responsible for __________ .

A. interference.

B. refraction.

C. reflection.

D. relativity.

36. A converging lens produces a real image ________________.

A. always.

B. never.

C. when the object is within one focal length of the lens.

D. when the object is further than one focal length from the lens.

37. The electromagnetic radiation with the longest wave length is/are _____________.

A. radio waves.

B. red light.

C. X-rays.

D. ultraviolet light.

38. Under a 440 power microscope, an object with diameter 0.1 millimeter appears to have a diameter of ___________ .

A. 4.4 millimeters.

B. 44 millimeters.

C. 440 millimeters.

D. 4400 millimeters.

39. Separating blood into blood cells and plasma involves the process of ___________ .

A. electrophoresis.

B. centrifugation.

C. spectrophotometry.

D. chromatography.

40. Experiments may be done with any of the following animals except ___________ .

A. birds.

B. invertebrates.

C. lower order life.

D. frogs.

41. For her first project of the year, a student is designing a science experiment to test the effects of light and water on plant growth. You should recommend that she ___________________.

A. manipulate the temperature also.

B. manipulate the water pH also.

C. determine the relationship between light and water unrelated to plant growth.

D. omit either water or light as a variable.

42. In a laboratory report, what is the abstract?

A. The abstract is a summary of the report, and is the first section of the report.

B. The abstract is a summary of the report, and is the last section of the report.

C. The abstract is predictions for future experiments, and is the first section of the report.

D. The abstract is predictions for future experiments, and is the last section of the report.

43. What is the scientific method?

A. It is the process of doing an experiment and writing a laboratory report.

B. It is the process of using open inquiry and repeatable results to establish theories.

C. It is the process of reinforcing scientific principles by confirming results.

D. It is the process of recording data and observations.

44. Identify the control in the following experiment: A student had four corn plants and was measuring photosynthetic rate (by measuring growth mass). Half of the plants were exposed to full (constant) sunlight, and the other half were kept in 50% (constant) sunlight.

A. The control is a set of plants grown in full (constant) sunlight.

B. The control is a set of plants grown in 50% (constant) sunlight.

B. The control is a set of plants grown in the dark.

D. The control is a set of plants grown in a mixture of natural levels of sunlight.

45. In an experiment measuring the growth of bacteria at different temperatures, what is the independent variable?

A. Number of bacteria.

B. Growth rate of bacteria.

C. Temperature.

D. Light intensity.

46. A scientific law__________.

A. proves scientific accuracy.

B. may never be broken.

C. may be revised in light of new data.

C. is the result of one excellent experiment.

47. Which is the correct order of methodology?

1. **collecting data**
2. **planning a controlled experiment**
3. **drawing a conclusion**
4. **hypothesizing a result**
5. **re-visiting a hypothesis to answer a question**

A. 1,2,3,4,5

B. 4,2,1,3,5

C. 4,5,1,3,2

D. 1,3,4,5,2

48. Which is the most desirable tool to use to heat substances in a middle school laboratory?

A. Alcohol burner.

B. Freestanding gas burner.

C. Bunsen burner.

D. Hot plate.

Duh!

49. Newton's Laws are taught in science classes because __________.

A. they are the correct analysis of inertia, gravity, and forces.

B. they are a close approximation to correct physics, for usual Earth conditions.

D. they accurately incorporate relativity into studies of forces.

E. Newton was a well-respected scientist in his time.

50. Which of the following is most accurate?

A. Mass is always constant; Weight may vary by location.

B. Mass and Weight are both always constant.

C. Weight is always constant; Mass may vary by location.

D. Mass and Weight may both vary by location.

51. Chemicals should be stored

A. in the principal's office.

B. in a dark room.

C. in an off-site research facility.

B. according to their reactivity with other substances.

52. Which of the following is the worst choice for a school laboratory activity?

A. A genetics experiment tracking the fur color of mice.

B. Dissection of a preserved fetal pig.

C. Measurement of goldfish respiration rate at different temperatures.

D. Pithing a frog to watch the circulatory system.

53. Who should be notified in the case of a serious chemical spill?

A. The custodian.

B. The fire department or their municipal authority.

C. The science department chair.

D. The School Board.

54. A scientist exposes mice to cigarette smoke, and notes that their lungs develop tumors. Mice that were not exposed to the smoke do not develop as many tumors. Which of the following conclusions may be drawn from these results?

I. Cigarette smoke causes lung tumors.
II. Cigarette smoke exposure has a positive correlation with lung tumors in mice.
III. Some mice are predisposed to develop lung tumors.
IV. Mice are often a good model for humans in scientific research.

A. I and II only.

B. II only.

C. I , II, and III only.

D. II and IV only.

55. In which situation would a science teacher be legally liable?

A. The teacher leaves the classroom for a telephone call and a student slips and injures him/herself.

B. A student removes his/her goggles and gets acid in his/her eye.

C. A faulty gas line in the classroom causes a fire.

D. A student cuts him/herself with a dissection scalpel.

56. Which of these is the best example of 'negligence'?

A. A teacher fails to give oral instructions to those with reading disabilities.

B. A teacher fails to exercise ordinary care to ensure safety in the classroom.

C. A teacher displays inability to supervise a large group of students.

D. A teacher reasonably anticipates that an event may occur, and plans accordingly.

57. Which item should always be used when handling glassware?

A. Tongs.

B. Safety goggles.

C. Gloves.

D. Buret stand.

58. Which of the following is *not* a necessary characteristic of living things?

Check this!

A. Movement.

B. Reduction of local entropy.

C. Ability to cause change in local energy form.

D. Reproduction.

59. What are the most significant and prevalent elements in the biosphere?

A. Carbon, Hydrogen, Oxygen, Nitrogen, Phosphorus.

B. Carbon, Hydrogen, Sodium, Iron, Calcium.

C. Carbon, Oxygen, Sulfur, Manganese, Iron.

D. Carbon, Hydrogen, Oxygen, Nickel, Sodium, Nitrogen.

60. All of the following measure energy *except* for __________

A. joules.

B. calories.

C. watts.

D. ergs.

61. Identify the correct sequence of organization of living things from lower to higher order:

A. Cell, Organelle, Organ, Tissue, System, Organism.

B. Cell, Tissue, Organ, Organelle, System, Organism.

C. Organelle, Cell, Tissue, Organ, System, Organism.

D. Organelle, Tissue, Cell, Organ, System, Organism.

62. Which kingdom is comprised of organisms made of one cell with no nuclear membrane?

A. Monera.

B. Protista.

C. Fungi.

D. Algae.

63. Which of the following is found in the least abundance in organic molecules?

A. Phosphorus.

B. Potassium.

C. Carbon.

D. Oxygen.

64. Catalysts assist reactions by __________ .

A. lowering effective activation energy.

B. maintaining precise pH levels.

C. keeping systems at equilibrium.

D. adjusting reaction speed.

65. Accepted procedures for preparing solutions should be made with __________ .

A. alcohol.

B. hydrochloric acid.

C. distilled water.

D. tap water.

66. Enzymes speed up reactions by __________ .

A. utilizing ATP.

B. lowering pH, allowing reaction speed to increase.

C. increasing volume of substrate.

D. lowering energy of activation.

67. When you step out of the shower, the floor feels colder on your feet than the bathmat. Which of the following is the correct explanation for this phenomenon?

A. The floor is colder than the bathmat.

B. Your feet have a chemical reaction with the floor, but not the bathmat.

C. Heat is conducted more easily into the floor.

D. Water is absorbed from your feet into the bathmat.

68. Which of the following is *not* considered ethical behavior for a scientist?

A. Using unpublished data and citing the source.

B. Publishing data before other scientists have had a chance to replicate results.

C. Collaborating with other scientists from different laboratories.

D. Publishing work with an incomplete list of citations.

69. The chemical equation for water formation is: $2H_2 + O_2 \rightarrow 2H_2O$. Which of the following is an *incorrect* interpretation of this equation?

A. Two moles of hydrogen gas and one mole of oxygen gas combine to make two moles of water.

B. Two grams of hydrogen gas and one gram of oxygen gas combine to make two grams of water.

C. Two molecules of hydrogen gas and one molecule of oxygen gas combine to make two molecules of water.

D. Four atoms of hydrogen (combined as a diatomic gas) and two atoms of oxygen (combined as a diatomic gas) combine to make two molecules of water.

70. Energy is measured with the same units as _____________.

A. force.

B. momentum.

C. work.

D. power.

71. If the volume of a confined gas is increased, what happens to the pressure of the gas? You may assume that the gas behaves ideally, and that temperature and number of gas molecules remain constant.

A. The pressure increases.

B. The pressure decreases.

C. The pressure stays the same.

D. There is not enough information given to answer this question.

72. A product of anaerobic respiration in animals is __________.

A. carbon dioxide.

B. lactic acid.

C. oxygen.

D. sodium chloride

73. A Newton is fundamentally a measure of __________ .

A. force.

B. momentum.

C. energy.

D. gravity.

74. Which change does *not* affect enzyme rate?

A. Increase the temperature.

B. Add more substrate.

C. Adjust the pH.

D. Use a larger cell.

75. Which of the following types of rock are made from magma?

A. Fossils

B. Sedimentary

C. Metamorphic

D. Igneous

76. Which of the following is *not* an acceptable way for a student to acknowledge sources in a laboratory report?

A. The student tells his/her teacher what sources s/he used to write the report.

B. The student uses footnotes in the text, with sources cited, but not in correct MLA format.

C. The student uses endnotes in the text, with sources cited, in correct MLA format.

D. The student attaches a separate bibliography, noting each use of sources.

77. Animals with a notochord or backbone are in the phylum

A. arthropoda.

B. chordata.

C. mollusca.

D. mammalia.

78. Which of the following is a correct explanation for scientific 'evolution'?

A. Giraffes need to reach higher for leaves to eat, so their necks stretch. The giraffe babies are then born with longer necks. Eventually, there are more long-necked giraffes in the population.

B. Giraffes with longer necks are able to reach more leaves, so they eat more and have more babies than other giraffes. Eventually, there are more long-necked giraffes in the population.

C. Giraffes want to reach higher for leaves to eat, so they release enzymes into their bloodstream, which in turn causes fetal development of longer-necked giraffes. Eventually, there are more long-necked giraffes in the population.

D. Giraffes with long necks are more attractive to other giraffes, so they get the best mating partners and have more babies. Eventually, there are more long-necked giraffes in the population.

79. Which of the following is a correct definition for 'chemical equilibrium'?

A. Chemical equilibrium is when the forward and backward reaction rates are equal. The reaction may continue to proceed forward and backward.

B. Chemical equilibrium is when the forward and backward reaction rates are equal, and equal to zero. The reaction does not continue.

C. Chemical equilibrium is when there are equal quantities of reactants and products.

D. Chemical equilibrium is when acids and bases neutralize each other fully.

80. Which of the following data sets is properly represented by a bar graph?

A. Number of people choosing to buy cars, vs. Color of car bought.

B. Number of people choosing to buy cars, vs. Age of car customer.

C. Number of people choosing to buy cars, vs. Distance from car lot to customer home.

D. Number of people choosing to buy cars, vs. Time since last car purchase.

81. In a science experiment, a student needs to dispense very small measured amounts of liquid into a well-mixed solution. Which of the following is the best choice for his/her equipment to use?

A. Buret with Buret Stand, Stir-plate, Stirring Rod, Beaker.

B. Buret with Buret Stand, Stir-plate, Beaker.

C. Volumetric Flask, Dropper, Graduated Cylinder, Stirring Rod.

D. Beaker, Graduated Cylinder, Stir-plate.

82. A laboratory balance is most appropriately used to measure the mass of which of the following?

A. Seven paper clips.

B. Three oranges.

C. Two hundred cells.

D. One student's elbow.

83. All of the following are measured in units of length, *except* for:

A. Perimeter.

B. Distance.

C. Radius.

D. Area.

84. What is specific gravity?

A. The mass of an object.

B. The ratio of the density of a substance to the density of water.

C. Density.

D. The pull of the earth's gravity on an object.

85. What is the most accurate description of the Water Cycle?

A. Rain comes from clouds, filling the ocean. The water then evaporates and becomes clouds again.

B. Water circulates from rivers into groundwater and back, while water vapor circulates in the atmosphere.

C. Water is conserved except for chemical or nuclear reactions, and any drop of water could circulate through clouds, rain, ground-water, and surface-water.

D. Weather systems cause chemical reactions to break water into its atoms.

86. The scientific name *Canis familiaris* refers to the animal's __________.

A. kingdom and phylum.

B. genus and species.

C. class and species.

D. type and family.

87. Members of the same animal species __________ .

A. look identical.

B. never adapt differently.

C. are able to reproduce with one another.

D. are found in the same location.

88. Which of the following is/are true about scientists?

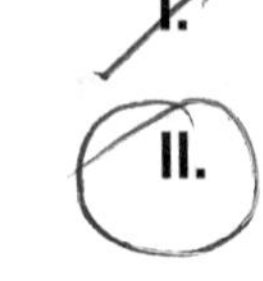

- **I. Scientists usually work alone.**
- **II. Scientists usually work with other scientists.**
- **III. Scientists achieve more prestige from new discoveries than from replicating established results.**
- **IV. Scientists keep records of their own work, but do not publish it for outside review.**

A. I and IV only.

B. II only.

C. II and III only.

D. I and IV only.

89. What is necessary for ion diffusion to occur spontaneously?

A. Carrier proteins.

B. Energy from an outside source.

C. A concentration gradient.

D. Cell flagellae.

90. All of the following are considered Newton's Laws *except* for:

A. An object in motion will continue in motion unless acted upon by an outside force.

B. For every action force, there is an equal and opposite reaction force.

C. Nature abhors a vacuum.

D. Mass can be considered the ratio of force to acceleration.

91. A cup of hot liquid and a cup of cold liquid are both sitting in a room at comfortable room temperature and humidity. Both cups are thin plastic. Which of the following is a true statement?

A. There will be fog on the outside of the hot liquid cup, and also fog on the outside of the cold liquid cup.

B. There will be fog on the outside of the hot liquid cup, but not on the cold liquid cup.

C. There will be fog on the outside of the cold liquid cup, but not on the hot liquid cup.

D. There will not be fog on the outside of either cup.

92. A ball rolls down a smooth hill. You may ignore air resistance. Which of the following is a true statement?

A. The ball has more energy at the start of its descent than just before it hits the bottom of the hill, because it is higher up at the beginning.

B. The ball has less energy at the start of its descent than just before it hits the bottom of the hill, because it is moving more quickly at the end.

C. The ball has the same energy throughout its descent, because positional energy is converted to energy of motion.

D. The ball has the same energy throughout its descent, because a single object (such as a ball) cannot gain or lose energy.

93. A long silver bar has a temperature of 50 degrees Celsius at one end and 0 degrees Celsius at the other end. The bar will reach thermal equilibrium (barring outside influence) by the process of heat _______________.

A. conduction.

B. convection.

C. radiation.

D. phase change.

94. ____________ are cracks in the plates of the earth's crust, along which the plates move.

A. Faults.

B. Ridges.

C. Earthquakes.

D. Volcanoes.

95. Fossils are usually found in _____________ rock.

A. igneous.

B. sedimentary.

C. metamorphic.

D. cumulus.

96. Which of the following is *not* a common type of acid in 'acid rain' or acidified surface water?

A. Nitric acid.

B. Sulfuric acid.

C. Carbonic acid.

D. Hydrofluoric acid.

97. Which of the following is *not* true about phase change in matter?

A. Solid water and liquid ice can coexist at water's freezing point.

B. At 7 degrees Celsius, water is always in liquid phase.

C. Matter changes phase when enough energy is gained or lost.

D. Different phases of matter are characterized by differences in molecular motion.

98. Which of the following is the longest (largest) unit of geological time?

A. Solar Year.

B. Epoch.

C. Period.

D. Era.

99. Extensive use of antibacterial soap has been found to increase the virulence of certain infections in hospitals. Which of the following might be an explanation for this phenomenon?

A. Antibacterial soaps do not kill viruses.

B. Antibacterial soaps do not incorporate the same antibiotics used as medicine.

C. Antibacterial soaps kill a lot of bacteria, and only the hardiest ones survive to reproduce.

D. Antibacterial soaps can be very drying to the skin.

100. Which of the following is a correct explanation for astronaut 'weightlessness'?

A. Astronauts continue to feel the pull of gravity in space, but they are so far from planets that the force is small.

B. Astronauts continue to feel the pull of gravity in space, but spacecraft have such powerful engines that those forces dominate, reducing effective weight.

C. Astronauts do not feel the pull of gravity in space, because space is a vacuum.

D. Astronauts do not feel the pull of gravity in space, because black hole forces dominate the force field, reducing their masses.

101. The theory of 'sea floor spreading' explains _________.

A. the shapes of the continents.

B. how continents collide.

C. how continents move apart.

D. how continents sink to become part of the ocean floor.

102. Which of the following animals are most likely to live in a tropical rain forest?

A. Reindeer.

B. Monkeys.

C. Puffins.

D. Bears.

103. Which of the following is *not* a type of volcano?

A. Shield Volcanoes.

B. Composite Volcanoes.

C. Stratus Volcanoes.

D. Cinder Cone Volcanoes.

104. Which of the following is *not* a property of metalloids?

A. Metalloids are solids at standard temperature and pressure.

B. Metalloids can conduct electricity to a limited extent.

C. Metalloids are found in groups 13 through 17.

D. Metalloids all favor ionic bonding.

105. Which of these is a true statement about loamy soil?

A. Loamy soil is gritty and porous.

B. Loamy soil is smooth and a good barrier to water.

C. Loamy soil is hostile to microorganisms.

D. Loamy soil is velvety and clumpy.

106. Lithification refers to the process by which unconsolidated sediments are transformed into ___________.

A. metamorphic rocks.

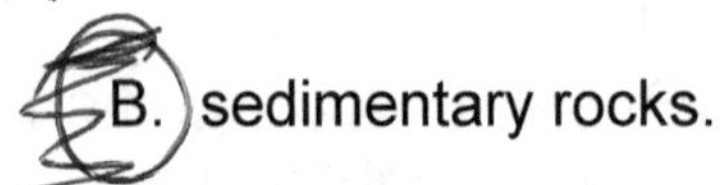

B. sedimentary rocks.

C. igneous rocks.

D. lithium oxide.

107. Igneous rocks can be classified according to which of the following?

A. Texture.

B. Composition.

C. Formation process.

D. All of the above.

108. Which of the following is the most accurate definition of a non-renewable resource?

A. A nonrenewable resource is never replaced once used.

B. A nonrenewable resource is replaced on a timescale that is very long relative to human life-spans.

C. A nonrenewable resource is a resource that can only be manufactured by humans.

D. A nonrenewable resource is a species that has already become extinct.

109. The theory of 'continental drift' is supported by which of the following?

A. The way the shapes of South America and Europe fit together.

B. The way the shapes of Europe and Asia fit together.

C. The way the shapes of South America and Africa fit together.

D. The way the shapes of North America and Antarctica fit together.

110. When water falls to a cave floor and evaporates, it may deposit calcium carbonate. This process leads to the formation of which of the following?

A. Stalactites.

B. Stalagmites.

C. Fault lines.

D. Sedimentary rocks.

111. A child has type O blood. Her father has type A blood, and her mother has type B blood. What are the genotypes of the father and mother, respectively?

A. AO and BO.

B. AA and AB.

C. OO and BO.

D. AO and BB.

112. Which of the following is the best definition for 'meteorite'?

A. A meteorite is a mineral composed of mica and feldspar.

B. A meteorite is material from outer space, that has struck the earth's surface.

C. A meteorite is an element that has properties of both metals and nonmetals.

D. A meteorite is a very small unit of length measurement.

113. A white flower is crossed with a red flower. Which of the following is a sign of incomplete dominance?

A. Pink flowers.

B. Red flowers.

C. White flowers.

D. No flowers.

114. What is the source for most of the United States' drinking water?

A. Desalinated ocean water.

B. Surface water (lakes, streams, mountain runoff).

C. Rainfall into municipal reservoirs.

D. Groundwater.

115. Which is the correct sequence of insect development?

A. Egg, pupa, larva, adult.

B. Egg, larva, pupa, adult.

C. Egg, adult, larva, pupa.

D. Pupa, egg, larva, adult.

116. A wrasse (fish) cleans the teeth of other fish by eating away plaque. This is an example of ___________ between the fish.

A. parasitism.

B. symbiosis (mutualism).

C. competition.

D. predation.

117. What is the main obstacle to using nuclear fusion for obtaining electricity?

A. Nuclear fusion produces much more pollution than nuclear fission.

B. There is no obstacle; most power plants us nuclear fusion today.

C. Nuclear fusion requires very high temperature and activation energy.

D. The fuel for nuclear fusion is extremely expensive.

118. Which of the following is a true statement about radiation exposure and air travel?

A. Air travel exposes humans to radiation, but the level is not significant for most people.

B. Air travel exposes humans to so much radiation that it is recommended as a method of cancer treatment.

C. Air travel does not expose humans to radiation.

D. Air travel may or may not expose humans to radiation, but it has not yet been determined.

119. Which process(es) result(s) in a haploid chromosome number?

A. Mitosis.

B. Meiosis.

C. Both mitosis and meiosis.

D. Neither mitosis nor meiosis.

120. Which of the following is *not* a member of Kingdom Fungi?

A. Mold.

B. Blue-green algae.

C. Mildew.

D. Mushrooms.

121. Which of the following organisms use spores to reproduce?

A. Fish.

B. Flowering plants.

C. Conifers.

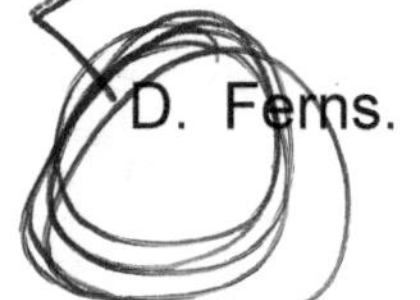

D. Ferns.

122. What is the main difference between the 'condensation hypothesis' and the 'tidal hypothesis' for the origin of the solar system?

A. The tidal hypothesis can be tested, but the condensation hypothesis cannot.

B. The tidal hypothesis proposes a near collision of two stars pulling on each other, but the condensation hypothesis proposes condensation of rotating clouds of dust and gas.

C. The tidal hypothesis explains how tides began on planets such as Earth, but the condensation hypothesis explains how water vapor became liquid on Earth.

D. The tidal hypothesis is based on Aristotelian physics, but the condensation hypothesis is based on Newtonian mechanics.

123. Which of the following units is *not* a measure of distance?

A. AU (astronomical unit).

B. Light year.

C. Parsec.

D. Lunar year.

124. The salinity of ocean water is closest to ____________ .

A. 0.035 %

B. 0.35 %

C. 3.5 %

D. 35 %

125. Which of the following will not change in a chemical reaction?

A. Number of moles of products.

B. Atomic number of one of the reactants.

C. Mass (in grams) of one of the reactants.

D. Rate of reaction.

Answer Key: Science

1.	B	26.	A	51.	D	76.	A	101.	C
2.	A	27.	B	52.	D	77.	B	102.	B
3.	A	28.	D	53.	B	78.	B	103.	C
4.	C	29.	B	54.	B	79.	A	104.	D
5.	B	30.	B	55.	A	80.	A	105.	D
6.	C	31.	C	56.	B	81.	B	106.	B
7.	D	32.	C	57.	B	82.	A	107.	D
8.	C	33.	D	58.	A	83.	D	108.	B
9.	C	34.	A	59.	A	84.	B	109.	C
10.	B	35.	B	60.	C	85.	C	110.	B
11.	A	36.	D	61.	C	86.	B	111.	A
12.	C	37.	A	62.	A	87.	C	112.	B
13.	C	38.	B	63.	B	88.	C	113.	A
14.	B	39.	C	64.	A	89.	C	114.	D
15.	B	40.	A	65.	C	90.	C	115.	B
16.	B	41.	D	66.	D	91.	C	116.	B
17.	A	42.	A	67.	C	92.	C	117.	C
18.	C	43.	B	68.	D	93.	A	118.	A
19.	C	44.	A	69.	B	94.	A	119.	B
20.	D	45.	C	70.	C	95.	B	120.	B
21.	C	46.	C	71.	B	96.	D	121.	D
22.	B	47.	B	72.	B	97.	B	122.	B
23.	A	48.	D	73.	A	98.	D	123.	D
24.	A	49.	B	74.	D	99.	C	124.	C
25.	A	50.	A	75.	D	100.	A	125.	B

Rationales with Sample Questions: Science

1. After an experiment, the scientist states that s/he believes a change in color is due to a change in pH. This is an example of

A. observing.

B. inferring.

C. measuring.

D. classifying.

B. Inferring.

To answer this question, note that the scientist has observed a change in color, and has then made a guess as to its reason. This is an example of inferring. The scientist has not measured or classified in this case. Although s/he has observed [the color change], the explanation of this observation is **inferring (B).**

2. When is a hypothesis formed?

A. Before the data is taken.

B. After the data is taken.

C. After the data is analyzed.

D. While the data is being graphed.

A. Before the data is taken.

A hypothesis is an educated guess, made before undertaking an experiment. The hypothesis is then evaluated based on the observed data. Therefore, the hypothesis must be formed before the data is taken, not during or after the experiment. This is consistent only with **answer (A).**

3. Who determines the laws regarding the use of safety glasses in the classroom?

A. The state government.

B. The school site.

C. The federal government.

D. The local district.

A. The state government.

Health and safety regulations are set by the state government, and apply to all school districts. Federal regulations may accompany specific federal grants, and local districts or school sites may enact local guidelines that are stricter than the state standards. All schools, however, must abide by safety precautions as set by state government. This is consistent only with **answer (A).**

4. If one inch equals 2.54 centimeters, how many millimeters are in 1.5 feet? (Approximately)

A. 18

B. 1800

C. 460

D. 4600

C. 460

To solve this problem, note that if one inch is 2.54 centimeters, then 1.5 feet (which is 18 inches), must be (18)(2.54) centimeters, i.e. approximately 46 centimeters. Because there are ten millimeters in a centimeter, this is approximately 460 millimeters:

(1.5 ft) (12 in/ft) (2.54 cm/in) (10 mm/cm) = (1.5) (12) (2.54) (10) mm = 457.2 mm

This is consistent only with **answer (C).**

5. Which of the following instruments measures wind speed?

A. Barometer.

B. Anemometer.

C. Thermometer.

D. Weather Vane.

B. Anemometer.

An anemometer is a device to measure wind speed, while a barometer measures pressure, a thermometer measures temperature, and a weather vane indicates wind direction. This is consistent only with **answer (B).**

If you chose "barometer," here is an old physics joke to console you:

A physics teacher asks a student the following question:
"Suppose you want to find out the height of a building, and the only tool you have is a barometer. How could you find out the height?"
(The teacher hopes that the student will remember that pressure is inversely proportional to height, and will measure the pressure at the top of the building and then use the data to calculate the height of the building.)
"Well," says the student, "I could tie a string to the barometer and lower it from the top of the building, and then measure the amount of string required."
"You could," answers the teacher, "but try to think of a method that uses your physics knowledge from our class."
"All right," replies the student, "I could drop the barometer from the roof and measure the time it takes to fall, and then use free-fall equations to calculate the height from which it fell."
"Yes," says the teacher, "but what about using the barometer per se?"
"Oh," answers the student, "I could find the building superintendent, and offer to exchange the barometer for a set of blueprints, and look up the height!"

6. Sonar works by

A. timing how long it takes sound to reach a certain speed.

B. bouncing sound waves between two metal plates.

C. bouncing sound waves off an object and timing how long it takes for the sound to return.

D. evaluating the motion and amplitude of sound.

C. Bouncing sound waves off an object and timing how long it takes for the sound to return.

Sonar is used to measure distances. Sound waves are sent out, and the time is measured for the sound to hit an obstacle and bounce back. By using the known speed of sound, observers (or machines) can calculate the distance to the obstacle. This is consistent only with **answer (C).**

7. The measure of the pull of Earth's gravity on an object is called

A. mass number.

B. atomic number.

C. mass.

D. weight.

D. Weight.

To answer this question, recall that mass number is the total number of protons and neutrons in an atom, atomic number is the number of protons in an atom, and mass is the amount of matter in an object. The only remaining **choice is (D)**, weight, which is correct because weight is the force of gravity on an object.

8. Which reaction below is a decomposition reaction?

A. $HCl + NaOH \rightarrow NaCl + H_2O$

B. $C + O_2 \rightarrow CO_2$

C. $2H_2O \rightarrow 2H_2 + O_2$

D. $CuSO_4 + Fe \rightarrow FeSO_4 + Cu$

C. $2H_2O \rightarrow 2H_2 + O_2$

To answer this question, recall that a decomposition reaction is one in which there are fewer reactants (on the left) than products (on the right). This is consistent only with **answer (C).** Meanwhile, note that answer (A) shows a double-replacement reaction (in which two sets of ions switch bonds), answer (B) shows a synthesis reaction (in which there are fewer products than reactants), and answer (D) shows a single-replacement reaction (in which one substance replaces another in its bond, but the other does not get a new bond).

9. The Law of Conservation of Energy states that

A. there must be the same number of products and reactants in any chemical equation.

B. objects always fall toward large masses such as planets.

C. energy is neither created nor destroyed, but may change form.

D. lights must be turned off when not in use, by state regulation.

C. Energy is neither created nor destroyed, but may change form.

Answer (C) is a summary of the Law of Conservation of Energy (for non-nuclear reactions). In other words, energy can be transformed into various forms such as kinetic, potential, electric, or heat energy, but the total amount of energy remains constant. Answer (A) is untrue, as demonstrated by many synthesis and decomposition reactions. Answers (B) and (D) may be sensible, but they are not relevant in this case. Therefore, the **answer is (C).**

10. Which parts of an atom are located inside the nucleus?

A. Protons and Electrons.

B. Protons and Neutrons.

C. Protons only.

D. Neutrons only.

B. Protons and Neutrons.

Protons and neutrons are located in the nucleus, while electrons move around outside the nucleus. This is consistent only with **answer (B)**.

11. The elements in the modern Periodic Table are arranged

A. in numerical order by atomic number.

B. randomly.

C. in alphabetical order by chemical symbol.

D. in numerical order by atomic mass.

A. In numerical order by atomic number.

Although the first periodic tables were arranged by atomic mass, the modern table is arranged by atomic number, i.e. the number of protons in each element. (This allows the element list to be complete and unique.) The elements are not arranged either randomly or in alphabetical order. The answer to this question is **therefore (A)**.

12. Carbon bonds with hydrogen by

A. ionic bonding.

B. non-polar covalent bonding.

C. polar covalent bonding.

D. strong nuclear force.

C. **Polar covalent bonding.**

Each carbon atom contains four valence electrons, while each hydrogen atom contains one valence electron. A carbon atom can bond with one or more hydrogen atoms, such that two electrons are shared in each bond. This is covalent bonding, because the electrons are shared. (In ionic bonding, atoms must gain or lose electrons to form ions. The ions are then electrically attracted in oppositely-charged pairs.) Covalent bonds are always polar when between two non-identical atoms, so this bond must be polar. ("Polar" means that the electrons are shared unequally, forming a pair of partial charges, i.e. poles.) In any case, the strong nuclear force is not relevant to this problem. The answer to this question is **therefore (C).**

13. Vinegar is an example of a

A. strong acid.

B. strong base.

C. weak acid.

D. weak base.

C. **Weak acid.**

The main ingredient in vinegar is acetic acid, a weak acid. Vinegar is a useful acid in science classes, because it makes a frothy reaction with bases such as baking soda (e.g. in the quintessential volcano model). Vinegar is not a strong acid, such as hydrochloric acid, because it does not dissociate as fully or cause as much corrosion. It is not a base. Therefore, the **answer is (C).**

14. Which of the following is not a nucleotide?

A. Adenine.

B. Alanine.

C. Cytosine.

D. Guanine.

B. Alanine.

Alanine is an amino acid. Adenine, cytosine, guanine, thymine, and uracil are nucleotides. The correct **answer is (B).**

15. When measuring the volume of water in a graduated cylinder, where does one read the measurement?

A. At the highest point of the liquid.

B. At the bottom of the meniscus curve.

C. At the closest mark to the top of the liquid.

D. At the top of the plastic safety ring.

B. At the bottom of the meniscus curve.

To measure water in glass, you must look at the top surface at eye-level, and ascertain the location of the bottom of the meniscus (the curved surface at the top of the water). The meniscus forms because water molecules adhere to the sides of the glass, which is a slightly stronger force than their cohesion to each other. This leads to a U-shaped top of the liquid column, the bottom of which gives the most accurate volume measurement. (Other liquids have different forces, e.g. mercury in glass, which has a convex meniscus.) This is consistent only with **answer (B).**

16. A duck's webbed feet are examples of

A. mimicry.

B. structural adaptation.

C. protective resemblance.

D. protective coloration.

B. Structural adaptation.

Ducks (and other aquatic birds) have webbed feet, which makes them more efficient swimmers. This is most likely due to evolutionary patterns where webbed-footed-birds were more successful at feeding and reproducing, and eventually became the majority of aquatic birds. Because the structure of the duck adapted to its environment over generations, this is termed 'structural adaptation'. Mimicry, protective resemblance, and protective coloration refer to other evolutionary mechanisms for survival. The answer to this question is **therefore (B)**.

17. What cell organelle contains the cell's stored food?

A. Vacuoles.

B. Golgi Apparatus.

C. Ribosomes.

D. Lysosomes.

A. Vacuoles.

In a cell, the sub-parts are called organelles. Of these, the vacuoles hold stored food (and water and pigments). The Golgi Apparatus sorts molecules from other parts of the cell; the ribosomes are sites of protein synthesis; the lysosomes contain digestive enzymes. This is consistent only with **answer (A).**

18. The first stage of mitosis is called

A. telophase.

B. anaphase.

C. prophase.

D. mitophase.

C. **Prophase.**

In mitosis, the division of somatic cells, prophase is the stage where the cell enters mitosis. The four stages of mitosis, in order, are: prophase, metaphase, anaphase, and telophase. ("Mitophase" is not one of the steps.) During prophase, the cell begins the nonstop process of division. Its chromatin condenses, its nucleolus disappears, the nuclear membrane breaks apart, mitotic spindles form, its cytoskeleton breaks down, and centrioles push the spindles apart. Note that interphase, the stage where chromatin is loose, chromosomes are replicated, and cell metabolism is occurring, is technically not a stage of mitosis; it is a precursor to cell division.

19. The Doppler Effect is associated most closely with which property of waves?

A. Amplitude.

B. Wavelength.

C. Frequency.

D. Intensity.

C. Frequency.

The Doppler Effect accounts for an apparent increase in frequency when a wave source moves toward a wave receiver or apparent decrease in frequency when a wave source moves away from a wave receiver. (Note that the receiver could also be moving toward or away from the source.) As the wave fronts are released, motion toward the receiver mimics more frequent wave fronts, while motion away from the receiver mimics less frequent wave fronts. Meanwhile, the amplitude, wavelength, and intensity of the wave are not as relevant to this process (although moving closer to a wave source makes it seem more intense). The **answer to this question is therefore (C)**.

20. Viruses are responsible for many human diseases including all of the following *except*

A. influenza.

B. A.I.D.S.

C. the common cold.

D. strep throat.

D. Strep throat.

Influenza, A.I.D.S., and the "common cold" (rhinovirus infection), are all caused by viruses. (This is the reason that doctors should not be pressured to prescribe antibiotics for colds or 'flu—i.e. they will not be effective since the infections are not bacterial.) Strep throat (properly called 'streptococcal throat' and caused by streptococcus bacteria) is not a virus, but a bacterial infection. Thus, the **answer is (D)**.

21. A series of experiments on pea plants formed by ________ showed that two invisible markers existed for each trait, and one marker dominated the other.

A. Pasteur.

B. Watson and Crick.

C. Mendel.

D. Mendeleev.

C. Mendel.

Gregor Mendel was a ninteenth-century Austrian botanist, who derived "laws" governing inherited traits. His work led to the understanding of dominant and recessive traits, carried by biological markers. Mendel cross-bred different kinds of pea plants with varying features and observed the resulting new plants. He showed that genetic characteristics are not passed identically from one generation to the next. (Pasteur, Watson, Crick, and Mendeleev were other scientists with different specialties.) This is consistent only with **answer (C)**.

22. Formaldehyde should not be used in school laboratories for the following reason:

A. it smells unpleasant.

B. it is a known carcinogen.

C. it is expensive to obtain.

D. it is an explosive.

B. It is a known carcinogen.

Formaldehyde is a known carcinogen, so it is too dangerous for use in schools. In general, teachers should not use carcinogens in school laboratories. Although formaldehyde also smells unpleasant, a smell alone is not a definitive marker of danger. For example, many people find the smell of vinegar to be unpleasant, but vinegar is considered a very safe classroom/laboratory chemical. Furthermore, some odorless materials are toxic. Formaldehyde is neither particularly expensive nor explosive. Thus, the **answer is (B)**.

23. Amino acids are carried to the ribosome in protein synthesis by:

A. transfer RNA (tRNA).

B. messenger RNA (mRNA).

C. ribosomal RNA (rRNA).

D. transformation RNA (trRNA).

A. Transfer RNA (tRNA).

The job of tRNA is to carry and position amino acids to/on the ribosomes. mRNA copies DNA code and brings it to the ribosomes; rRNA is in the ribosome itself. There is no such thing as trRNA. Thus, the **answer is (A)**.

24. When designing a scientific experiment, a student considers all the factors that may influence the results. The process goal is to

A. recognize and manipulate independent variables.

B. recognize and record independent variables.

C. recognize and manipulate dependent variables.

D. recognize and record dependent variables.

A. Recognize and manipulate independent variables.

When a student designs a scientific experiment, s/he must decide what to measure, and what independent variables will play a role in the experiment. S/he must determine how to manipulate these independent variables to refine his/her procedure and to prepare for meaningful observations. Although s/he will eventually record dependent variables (D), this does not take place during the experimental design phase. Although the student will likely recognize and record the independent variables (B), this is not the process goal, but a helpful step in manipulating the variables. It is unlikely that the student will manipulate dependent variables directly in his/her experiment (C), or the data would be suspect. Thus, the **answer is (A)**.

25. Since ancient times, people have been entranced with bird flight. What is the key to bird flight?

A. Bird wings are a particular shape and composition.

B. Birds flap their wings quickly enough to propel themselves.

C. Birds take advantage of tailwinds.

D. Birds take advantage of crosswinds.

A. Bird wings are a particular shape and composition.

Bird wings are shaped for wide area, and their bones are very light. This creates a large surface-area-to-mass ratio, enabling birds to glide in air. Birds do flap their wings and float on winds, but none of these is the main reason for their flight ability. Thus, the **answer is (A)**.

26. Laboratory researchers have classified fungi as distinct from plants because the cell walls of fungi

A. contain chitin.

B. contain yeast.

C. are more solid.

D. are less solid.

A. Contain chitin.

Kingdom Fungi consists of organisms that are eukaryotic, multicellular, absorptive consumers. They have a chitin cell wall, which is the only universally present feature in fungi that is never present in plants. Thus, the **answer is (A)**.

27. In a fission reactor, "heavy water" is used to

A. terminate fission reactions.

B. slow down neutrons and moderate reactions.

C. rehydrate the chemicals.

D. initiate a chain reaction.

B. Slow down neutrons and moderate reactions.

"Heavy water" is used in a nuclear [fission] reactor to slow down neutrons, controlling and moderating the nuclear reactions. It does not terminate the reaction, and it does not initiate the reaction. Also, although the reactor takes advantage of water's other properties (e.g. high specific heat for cooling), the water does not "rehydrate" the chemicals. Therefore, the **answer is (B)**.

28. The transfer of heat by electromagnetic waves is called

A. conduction.

B. convection.

C. phase change.

D. radiation.

D. Radiation.

Heat transfer via electromagnetic waves (which can occur even in a vacuum) is called radiation. (Heat can also be transferred by direct contact (conduction), by fluid current (convection), and by matter changing phase, but these are not relevant here.) The answer to this question is **therefore (D)**.

29. When heat is added to most solids, they expand. Why is this the case?

A. The molecules get bigger.

B. The faster molecular motion leads to greater distance between the molecules.

C. The molecules develop greater repelling electric forces.

D. The molecules form a more rigid structure.

B. The faster molecular motion leads to greater distance between the molecules.

The atomic theory of matter states that matter is made up of tiny, rapidly moving particles. These particles move more quickly when warmer, because temperature is a measure of average kinetic energy of the particles. Warmer molecules therefore move further away from each other, with enough energy to separate from each other more often and for greater distances. The individual molecules do not get bigger, by conservation of mass, eliminating answer (A). The molecules do not develop greater repelling electric forces, eliminating answer (C). Occasionally, molecules form a more rigid structure when becoming colder and freezing (such as water)—but this gives rise to the exceptions to heat expansion, so it is not relevant here, eliminating answer (D). Therefore, the **answer is (B)**.

30. The force of gravity on earth causes all bodies in free fall to

A. fall at the same speed.

B. accelerate at the same rate.

C. reach the same terminal velocity.

D. move in the same direction.

B. Accelerate at the same rate.

Gravity causes approximately the same acceleration on all falling bodies close to earth's surface. (It is only "approximately" because there are very small variations in the strength of earth's gravitational field.) More massive bodies continue to accelerate at this rate for longer, before their air resistance is great enough to cause terminal velocity, so answers (A) and (C) are eliminated. Bodies on different parts of the planet move in different directions (always toward the center of mass of earth), so answer (D) is eliminated. Thus, the **answer is (B)**.

31. Sound waves are produced by

A. pitch.

B. noise.

C. vibrations.

D. sonar.

C. Vibrations.

Sound waves are produced by a vibrating body. The vibrating object moves forward and compresses the air in front of it, then reverses direction so that the pressure on the air is lessened and expansion of the air molecules occurs. The vibrating air molecules move back and forth parallel to the direction of motion of the wave as they pass the energy from adjacent air molecules closer to the source to air molecules farther away from the source. Therefore, the **answer is (C)**.

32. Resistance is measured in units called

A. watts.

B. volts.

C. ohms.

D. current.

C. Ohms.

A watt is a unit of energy. Potential difference is measured in a unit called the volt. Current is the number of electrons per second that flow past a point in a circuit. An ohm is the unit for resistance. The correct **answer is (C).**

33. Sound can be transmitted in all of the following *except*

A. air.

B. water.

C. diamond.

D. a vacuum.

D. A vacuum.

Sound, a longitudinal wave, is transmitted by vibrations of molecules. Therefore, it can be transmitted through any gas, liquid, or solid. However, it cannot be transmitted through a vacuum, because there are no particles present to vibrate and bump into their adjacent particles to transmit the waves. This is consistent only with **answer (D)**. (It is interesting also to note that sound is actually faster in solids and liquids than in air.)

34. As a train approaches, the whistle sounds

A. higher, because it has a higher apparent frequency.

B. lower, because it has a lower apparent frequency.

C. higher, because it has a lower apparent frequency.

D. lower, because it has a higher apparent frequency.

A. Higher, because it has a higher apparent frequency.

By the Doppler effect, when a source of sound is moving toward an observer, the wave fronts are released closer together, i.e. with a greater apparent frequency. Higher frequency sounds are higher in pitch. This is consistent only with **answer (A)**.

35. The speed of light is different in different materials. This is responsible for

A. interference.

B. refraction.

C. reflection.

D. relativity.

B. Refraction.

Refraction (B) is the bending of light because it hits a material at an angle wherein it has a different speed. (This is analogous to a cart rolling on a smooth road. If it hits a rough patch at an angle, the wheel on the rough patch slows down first, leading to a change in direction.) Interference (A) is when light waves interfere with each other to form brighter or dimmer patterns; reflection (C) is when light bounces off a surface; relativity (D) is a general topic related to light speed and its implications, but not specifically indicated here. Therefore, the **answer is (B)**.

36. A converging lens produces a real image _______________

A. always.

B. never.

C. when the object is within one focal length of the lens.

D. when the object is further than one focal length from the lens.

D. When the object is further than one focal length from the lens.

A converging lens produces a real image whenever the object is far enough from the lens (outside one focal length) so that the rays of light from the object can hit the lens and be focused into a real image on the other side of the lens. When the object is closer than one focal length from the lens, rays of light do not converge on the other side; they diverge. This means that only a virtual image can be formed, i.e. the theoretical place where those diverging rays would have converged if they had originated behind the object. Thus, the correct **answer is (D)**.

37. The electromagnetic radiation with the longest wave length is ____________

A. radio waves.

B. red light.

C. X-rays.

D. ultraviolet light.

A. Radio waves.

As one can see on a diagram of the electromagnetic spectrum, radio waves have longer wave lengths (and smaller frequencies) than visible light, which in turn has longer wave lengths than ultraviolet or X-ray radiation. If you did not remember this sequence, you might recall that wave length is inversely proportional to frequency, and that radio waves are considered much less harmful (less energetic, i.e. lower frequency) than ultraviolet or X-ray radiation. The correct answer is **therefore (A)**.

38. Under a 440 power microscope, an object with diameter 0.1 millimeter appears to have diameter _____________

A. 4.4 millimeters.

B. 44 millimeters.

C. 440 millimeters.

D. 4400 millimeters.

B. 44 millimeters.

To answer this question, recall that to calculate a new length, you multiply the original length by the magnification power of the instrument. Therefore, the 0.1 millimeter diameter is multiplied by 440. This equals 44, so the image appears to be 44 millimeters in diameter. You could also reason that since a 440 power microscope is considered a "high power" microscope, you would expect a 0.1 millimeter object to appear a few centimeters long. Therefore, the correct **answer is (B)**.

39. To separate blood into blood cells and plasma involves the process of

A. electrophoresis.

B. centrifugation.

C. spectrophotometry.

D. chromatography.

C. Centrifugation.

Electrophoresis uses electrical charges of molecules to separate them according to their size. Spectrophotometry uses percent light absorbance to measure a color change, thus giving qualitative data a quantitative value. Chromatography uses the principles of capillarity to separate substances. Centrifugation involves spinning substances at a high speed. The more dense part of a solution will settle to the bottom of the test tube, where the lighter material will stay on top. The **answer is (C).**

40. Experiments may be done with any of the following animals except

A. birds.

B. invertebrates.

C .lower order life.

D. frogs.

A. Birds.

No dissections may be performed on living mammalian vertebrates or birds. Lower order life and invertebrates may be used. Biological experiments may be done with all animals except mammalian vertebrates or birds. Therefore the **answer is (A).**

41. For her first project of the year, a student is designing a science experiment to test the effects of light and water on plant growth. You should recommend that she ________________

A. manipulate the temperature also.

B. manipulate the water pH also.

C. determine the relationship between light and water unrelated to plant growth.

D. omit either water or light as a variable.

D. **Omit either water or light as a variable.**

As a science teacher for middle-school-aged kids, it is important to reinforce the idea of 'constant' vs. 'variable' in science experiments. At this level, it is wisest to have only one variable examined in each science experiment. (Later, students can hold different variables constant while investigating others.) Therefore it is counterproductive to add in other variables (answers (A)or (B)). It is also irrelevant to determine the light-water interactions aside from plant growth (C). So the only possible **answer is (D)**.

42. In a laboratory report, what is the abstract?

A. The abstract is a summary of the report, and is the first section of the report.

B. The abstract is a summary of the report, and is the last section of the report.

C. The abstract is predictions for future experiments, and is the first section of the report.

D. The abstract is predictions for future experiments, and is the last section of the report.

A. The abstract is a summary of the report, and is the first section of the report.

In a laboratory report, the abstract is the section that summarizes the entire report (often containing one representative sentence from each section). It appears at the very beginning of the report, even before the introduction, often on its own page (instead of a title page). This format is consistent with articles in scientific journals. Therefore, the **answer is (A).**

43. What is the scientific method?

A. It is the process of doing an experiment and writing a laboratory report.

B. It is the process of using open inquiry and repeatable results to establish theories.

C. It is the process of reinforcing scientific principles by confirming results.

D. It is the process of recording data and observations.

B. It is the process of using open inquiry and repeatable results to establish theories.

Scientific research often includes elements from answers (A), (C), and (D), but the basic underlying principle of the scientific method is that people ask questions and do repeatable experiments to answer those questions and develop informed theories of why and how things happen. Therefore, the best **answer is (B).**

44. Identify the control in the following experiment: A student had four corn plants and was measuring photosynthetic rate (by measuring growth mass). Half of the plants were exposed to full (constant) sunlight, and the other half were kept in 50% (constant) sunlight.

A. The control is a set of plants grown in full (constant) sunlight.

B. The control is a set of plants grown in 50% (constant) sunlight.

C. The control is a set of plants grown in the dark.

D. The control is a set of plants grown in a mixture of natural levels of sunlight.

A. The control is a set of plants grown in full (constant) sunlight.

In this experiment, the goal was to measure how two different amounts of sunlight affected plant growth. The control in any experiment is the 'base case,' or the usual situation without a change in variable. Because the control must be studied alongside the variable, answers (C) and (D) are omitted (because they were not in the experiment). The **better answer of (A) and (B) is (A)**, because usually plants are assumed to have the best growth and their usual growing circumstances in full sunlight. This is particularly true for crops like the corn plants in this question.

45. In an experiment measuring the growth of bacteria at different temperatures, what is the independent variable?

A. Number of bacteria.

B. Growth rate of bacteria.

C. Temperature.

D. Light intensity.

C. Temperature.

To answer this question, recall that the independent variable in an experiment is the entity that is changed by the scientist, in order to observe the effects (the dependent variable(s)). In this experiment, temperature is changed in order to measure growth of bacteria, so **(C) is the answer**. Note that answer (A) is the dependent variable, and neither (B) nor (D) is directly relevant to the question.

46. A scientific law _____________

A. proves scientific accuracy.

B. may never be broken.

C. may be revised in light of new data.

D. is the result of one excellent experiment.

C. **May be revised in light of new data.**

A scientific law is the same as a scientific theory, except that it has lasted for longer, and has been supported by more extensive data. Therefore, such a law may be revised in light of new data, and may be broken by that new data. Furthermore, a scientific law is always the result of many experiments, and never 'proves' anything but rather is implied or supported by various results. Therefore, the **answer must be (C).**

47. Which is the correct order of methodology?

6. **collecting data**
7. **planning a controlled experiment**
8. **drawing a conclusion**
9. **hypothesizing a result**
10. **re-visiting a hypothesis to answer a question**

A. 1,2,3,4,5

B. 4,2,1,3,5

C. 4,5,1,3,2

D. 1,3,4,5,2

B. **4.2.1.3.5**

The correct methodology for the scientific method is first to make a meaningful hypothesis (educated guess), then plan and execute a controlled experiment to test that hypothesis. Using the data collected in that experiment, the scientist then draws conclusions and attempts to answer the original question related to the hypothesis. This is consistent only with **answer (B)**.

48. Which is the most desirable tool to use to heat substances in a middle school laboratory?

A. Alcohol burner.

B. Freestanding gas burner.

C. Bunsen burner.

D. Hot plate.

D. Hot plate.

Due to safety considerations, the use of open flame should be minimized, so a hot plate is the best choice. Any kind of burner may be used with proper precautions, but it is difficult to maintain a completely safe middle school environment. Therefore, the best **answer is (D)**.

49. Newton's Laws are taught in science classes because __________________

A. they are the correct analysis of inertia, gravity, and forces.

B. they are a close approximation to correct physics, for usual Earth conditions.

C. they accurately incorporate Relativity into studies of forces.

D. Newton was a well-respected scientist in his time.

B. They are a close approximation to correct physics, for usual Earth conditions.

Although Newton's Laws are often taught as fully correct for inertia, gravity, and forces, it is important to realize that Einstein's work (and that of others) has indicated that Newton's Laws are reliable only at speeds much lower than that of light. This is reasonable, though, for most middle- and high-school applications. At speeds close to the speed of light, Relativity considerations must be used. Therefore, the only correct **answer is (B)**.

50. Which of the following is most accurate?

A. Mass is always constant; Weight may vary by location.

B. Mass and Weight are both always constant.

C. Weight is always constant; Mass may vary by location.

D. Mass and Weight may both vary by location.

A. **Mass is always constant; Weight may vary by location**.

When considering situations exclusive of nuclear reactions, mass is constant (mass, the amount of matter in a system, is conserved). Weight, on the other hand, is the force of gravity on an object, which is subject to change due to changes in the gravitational field and/or the location of the object. Thus, the **best answer is (A)**.

51. Chemicals should be stored __________________

A. in the principal's office.

B. in a dark room.

C. in an off-site research facility.

D. according to their reactivity with other substances.

D. **According to their reactivity with other substances.**

Chemicals should be stored with other chemicals of similar properties (e.g. acids with other acids), to reduce the potential for either hazardous reactions in the store-room, or mistakes in reagent use. Certainly, chemicals should not be stored in anyone's office, and the light intensity of the room is not very important because light-sensitive chemicals are usually stored in dark containers. In fact, good lighting is desirable in a store-room, so that labels can be read easily. Chemicals may be stored off-site, but that makes their use inconvenient. Therefore, the best **answer is (D)**.

52. Which of the following is the worst choice for a school laboratory activity?

A. A genetics experiment tracking the fur color of mice.

B. Dissection of a preserved fetal pig.

C. Measurement of goldfish respiration rate at different temperatures.

D. Pithing a frog to watch the circulatory system.

D. Pithing a frog to watch the circulatory system.

While any use of animals (alive or dead) must be done with care to respect ethics and laws, it is possible to perform choices (A), (B), or (C) with due care. (Note that students will need significant assistance and maturity to perform these experiments.) However, modern practice precludes pithing animals (causing partial brain death while allowing some systems to function), as inhumane. Therefore, the answer to this **question is (D)**.

53. Who should be notified in the case of a serious chemical spill?

A. The custodian.

B. The fire department or other municipal authority.

C. The science department chair.

D. The School Board.

B. The fire department or other municipal authority.

Although the custodian may help to clean up laboratory messes, and the science department chair should be involved in discussions of ways to avoid spills, a serious chemical spill may require action by the fire department or other trained emergency personnel. It is best to be safe by notifying them in case of a serious chemical accident. Therefore, the **best answer is (B)**.

54. A scientist exposes mice to cigarette smoke, and notes that their lungs develop tumors. Mice that were not exposed to the smoke do not develop as many tumors. Which of the following conclusions may be drawn from these results?

I. Cigarette smoke causes lung tumors.
II. Cigarette smoke exposure has a positive correlation with lung tumors in mice.
III. Some mice are predisposed to develop lung tumors.
IV. Mice are often a good model for humans in scientific research.

A. I and II only.

B. II only.

C. I , II, and III only.

D. II and IV only.

B. II only.

Although cigarette smoke has been found to cause lung tumors (and many other problems), this particular experiment shows only that there is a positive correlation between smoke exposure and tumor development in these mice. It may be true that some mice are more likely to develop tumors than others, which is why a control group of identical mice should have been used for comparison. Mice are often used to model human reactions, but this is as much due to their low financial and emotional cost as it is due to their being a "good model" for humans. Therefore, the **answer must be (B)**.

55. In which situation would a science teacher be legally liable?

A. The teacher leaves the classroom for a telephone call and a student slips and injures him/herself.

B. A student removes his/her goggles and gets acid in his/her eye.

C. A faulty gas line in the classroom causes a fire.

D. A student cuts him/herself with a dissection scalpel.

A. The teacher leaves the classroom for a telephone call and a student slips and injures him/herself.

Teachers are required to exercise a "reasonable duty of care" for their students. Accidents may happen (e.g. (D)), or students may make poor decisions (e.g. (B)), or facilities may break down (e.g. (C)). However, the teacher has the responsibility to be present and to do his/her best to create a safe and effective learning environment. Therefore, the **answer is (A)**.

56. Which of these is the best example of 'negligence'?

A. A teacher fails to give oral instructions to those with reading disabilities.

B. A teacher fails to exercise ordinary care to ensure safety in the classroom.

C. A teacher displays inability to supervise a large group of students.

D. A teacher reasonably anticipates that an event may occur, and plans accordingly.

B. A teacher fails to exercise ordinary care to ensure safety in the classroom.

'Negligence' is the failure to "exercise ordinary care" to ensure an appropriate and safe classroom environment. It is best for a teacher to meet all special requirements for disabled students, and to be good at supervising large groups. However, if a teacher can prove that s/he has done a reasonable job to ensure a safe and effective learning environment, then it is unlikely that she/he would be found negligent. Therefore, **the answer is (B)**.

57. Which item should always be used when handling glassware?

A. Tongs.

B. Safety goggles.

C. Gloves.

D. Buret stand.

B. Safety goggles.

Safety goggles are the single most important piece of safety equipment in the laboratory, and should be used any time a scientist is using glassware, heat, or chemicals. Other equipment (e.g. tongs, gloves, or even a buret stand) has its place for various applications. However, the most important is safety goggles. Therefore, the **answer is (B)**.

58. Which of the following is *not* a necessary characteristic of living things?

A. Movement.

B. Reduction of local entropy.

C. Ability to cause local energy form changes.

D. Reproduction.

A. Movement.

There are many definitions of "life," but in all cases, a living organism reduces local entropy, changes chemical energy into other forms, and reproduces. Not all living things move, however, so the correct **answer is (A)**.

59. What are the most significant and prevalent elements in the biosphere?

A. Carbon, Hydrogen, Oxygen, Nitrogen, Phosphorus.

B. Carbon, Hydrogen, Sodium, Iron, Calcium.

C. Carbon, Oxygen, Sulfur, Manganese, Iron.

D. Carbon, Hydrogen, Oxygen, Nickel, Sodium, Nitrogen.

A. Carbon, Hydrogen, Oxygen, Nitrogen, Phosphorus.

Organic matter (and life as we know it) is based on Carbon atoms, bonded to Hydrogen and Oxygen. Nitrogen and Phosphorus are the next most significant elements, followed by Sulfur and then trace nutrients such as Iron, Sodium, Calcium, and others. Therefore, the **answer is (A)**. If you know that the formula for any carbohydrate contains Carbon, Hydrogen, and Oxygen, that will help you narrow the choices to (A) and (D) in any case.

60. All of the following measure energy *except* for __________

A. joules.

B. calories.

C. watts.

D. ergs.

C. Watts.

Energy units must be dimensionally equivalent to (force)x(length), which equals (mass)x(length squared)/(time squared). Joules, Calories, and Ergs are all metric measures of energy. Joules are the SI units of energy, while Calories are used to allow water to have a Specific Heat of one unit. Ergs are used in the 'cgs' (centimeter-gram-second) system, for smaller quantities. Watts, however, are units of power, i.e. Joules per Second. Therefore, the **answer is (C)**.

61. Identify the correct sequence of organization of living things from lower to higher order:

A. Cell, Organelle, Organ, Tissue, System, Organism.

B. Cell, Tissue, Organ, Organelle, System, Organism.

C. Organelle, Cell, Tissue, Organ, System, Organism.

D. Organelle, Tissue, Cell, Organ, System, Organism.

C. **Organelle, Cell, Tissue, Organ, System, Organism**.

Organelles are parts of the cell; cells make up tissue, which makes up organs. Organs work together in systems (e.g. the respiratory system), and the organism is the living thing as a whole. Therefore, the **answer must be (C)**.

62. Which kingdom is comprised of organisms made of one cell with no nuclear membrane?

A. Monera.

B. Protista.

C. Fungi.

D. Algae.

A. **Monera.**

To answer this question, first note that algae are not a kingdom of their own. Some algae are in monera, the kingdom that consists of unicellular prokaryotes with no true nucleus. Protista and fungi are both eukaryotic, with true nuclei, and are sometimes multi-cellular. Therefore, the **answer is (A)**.

63. Which of the following is found in the least abundance in organic molecules?

A. Phosphorus.

B. Potassium.

C. Carbon.

D. Oxygen.

B. Potassium.

Organic molecules consist mainly of Carbon, Hydrogen, and Oxygen, with significant amounts of Nitrogen, Phosphorus, and often Sulfur. Other elements, such as Potassium, are present in much smaller quantities. Therefore, the **answer is (B)**. If you were not aware of this ranking, you might have been able to eliminate Carbon and Oxygen because of their prevalence, in any case.

64. Catalysts assist reactions by ______________

A. lowering effective activation energy.

B. maintaining precise pH levels.

C. keeping systems at equilibrium.

D. adjusting reaction speed.

A. Lowering effective activation energy.

Chemical reactions can be enhanced or accelerated by catalysts, which are present both with reactants and with products. They induce the formation of activated complexes, thereby lowering the effective activation energy—so that less energy is necessary for the reaction to begin. Although this often makes reactions faster, answer (D) is not as good a choice as the more generally applicable **answer (A)**, which is correct.

65. Accepted procedures for preparing solutions should be made with

A. alcohol.

B. hydrochloric acid.

C. distilled water.

D. tap water.

C. Distilled water.

Alcohol and hydrochloric acid should never be used to make solutions unless instructed to do so. All solutions should be made with distilled water as tap water contains dissolved particles which may affect the results of an experiment. The correct **answer is (C).**

66. Enzymes speed up reactions by ______________

A. utilizing ATP.

B. lowering pH, allowing reaction speed to increase.

C. increasing volume of substrate.

D. lowering energy of activation.

D. **Lowering energy of activation**.

Because enzymes are catalysts, they work the same way—they cause the formation of activated chemical complexes, which require a lower activation energy. Therefore, the **answer is (D).** ATP is an energy source for cells, and pH or volume changes may or may not affect reaction rate, so these answers can be eliminated.

67. When you step out of the shower, the floor feels colder on your feet than the bathmat. Which of the following is the correct explanation for this phenomenon?

A. The floor is colder than the bathmat.

B. Your feet have a chemical reaction with the floor, but not the bathmat.

C. Heat is conducted more easily into the floor.

D. Water is absorbed from your feet into the bathmat.

C. Heat is conducted more easily into the floor.

When you step out of the shower and onto a surface, the surface is most likely at room temperature, regardless of its composition (eliminating answer (A)). Your feet feel cold when heat is transferred from them to the surface, which happens more easily on a hard floor than a soft bathmat. This is because of differences in specific heat (the energy required to change temperature, which varies by material). Therefore, the **answer must be (C)**, i.e. heat is conducted more easily into the floor from your feet.

68. Which of the following is *not* considered ethical behavior for a scientist?

A. Using unpublished data and citing the source.

B. Publishing data before other scientists have had a chance to replicate results.

C. Collaborating with other scientists from different laboratories.

D. Publishing work with an incomplete list of citations.

D. Publishing work with an incomplete list of citations.

One of the most important ethical principles for scientists is to cite all sources of data and analysis when publishing work. It is reasonable to use unpublished data (A), as long as the source is cited. Most science is published before other scientists replicate it (B), and frequently scientists collaborate with each other, in the same or different laboratories (C). These are all ethical choices. However, publishing work without the appropriate citations, is unethical. Therefore, the **answer is (D).**

69. The chemical equation for water formation is: $2H_2 + O_2 \rightarrow 2H_2O$. Which of the following is an *incorrect* interpretation of this equation?

A. Two moles of hydrogen gas and one mole of oxygen gas combine to make two moles of water.

B. Two grams of hydrogen gas and one gram of oxygen gas combine to make two grams of water.

C. Two molecules of hydrogen gas and one molecule of oxygen gas combine to make two molecules of water.

D. Four atoms of hydrogen (combined as a diatomic gas) and two atoms of oxygen (combined as a diatomic gas) combine to make two molecules of water.

B. Two grams of hydrogen gas and one gram of oxygen gas combine to make two grams of water.

In any chemical equation, the coefficients indicate the relative proportions of molecules (or atoms), or of moles of molecules. They do not refer to mass, because chemicals combine in repeatable combinations of molar ratio (i.e. number of moles), but vary in mass per mole of material. Therefore, the answer must be the only choice that does not refer to numbers of particles, i.e. **answer (B)**, which refers to grams, a unit of mass.

70. Energy is measured with the same units as ______________

A. force.

B. momentum.

C. work.

D. power.

C. Work.

In SI units, energy is measured in Joules, i.e. (mass)(length squared)/(time squared). This is the same unit as is used for work. You can verify this by calculating that since work is force times distance, the units work out to be the same. Force is measured in Newtons in SI; momentum is measured in (mass)(length)/(time); power is measured in Watts (which equal Joules/second). Therefore, the **answer must be (C)**.

71. If the volume of a confined gas is increased, what happens to the pressure of the gas? You may assume that the gas behaves ideally, and that temperature and number of gas molecules remain constant.

A. The pressure increases.

B. The pressure decreases.

C. The pressure stays the same.

D. There is not enough information given to answer this question.

B. The pressure decreases.

Because we are told that the gas behaves ideally, you may assume that it follows the Ideal Gas Law, i.e. $PV = nRT$. This means that an increase in volume must be associated with a decrease in pressure (i.e. higher T means lower P), because we are also given that all the components of the right side of the equation remain constant. Therefore, the **answer must be (B)**.

72. A product of anaerobic respiration in animals is ______________

A. carbon dioxide.

B. lactic acid.

C. oxygen.

D. sodium chloride.

B. Lactic acid.

In animals, anaerobic respiration (i.e. respiration without the presence of oxygen) generates lactic acid as a byproduct. (Note that some anaerobic bacteria generate carbon dioxide from respiration of methane, and animals generate carbon dioxide in aerobic respiration.) Oxygen is not normally a by-product of respiration, though it is a product of photosynthesis, and sodium chloride is not strictly relevant in this question. Therefore, the **answer must be (B)**. By the way, lactic acid is believed to cause muscle soreness after anaerobic weight-lifting.

73. A Newton is fundamentally a measure of ______________

A. force.

B. momentum.

C. energy.

D. gravity.

A. Force.

In SI units, force is measured in Newtons. Momentum and energy each have different units, without equivalent dimensions. A Newton is one (kilogram)(meter)/(second squared), while momentum is measured in (kilgram)(meter)/(second) and energy, in Joules, is (kilogram)(meter squared)/(second squared). Although "gravity" can be interpreted as the force of gravity, i.e. measured in Newtons, fundamentally it is not required. Therefore, the **answer is (A)**.

74. Which change does *not* affect enzyme rate?

A. Increase the temperature.

B. Add more substrate.

C. Adjust the pH.

D. Use a larger cell.

D. Use a larger cell.

Temperature, chemical amounts, and pH can all affect enzyme rate. However, the chemical reactions take place on a small enough scale that the overall cell size is not relevant. Therefore, the **answer is (D)**.

75. Which of the following types of rock are made from magma?

A. Fossils.

B. Sedimentary.

C. Metamorphic.

D. Igneous.

D. Igneous.

Few fossils are found in metamorphic rock and virtually none found in igneous rocks. Igneous rocks are formed from magma and magma is so hot that any organisms trapped by it are destroyed. Metamorphic rocks are formed by high temperatures and great pressures. When fluid sediments are transformed into solid sedimentary rocks, the process is known as lithification. The **answer is (D).**

76. Which of the following is *not* an acceptable way for a student to acknowledge sources in a laboratory report?

A. The student tells his/her teacher what sources s/he used to write the report.

B. The student uses footnotes in the text, with sources cited, but not in correct MLA format.

C. The student uses endnotes in the text, with sources cited, in correct MLA format.

D. The student attaches a separate bibliography, noting each use of sources.

A. The student tells his/her teacher what sources s/he used to write the report.

It may seem obvious, but students are often unaware that scientists need to cite all sources used. For the young adolescent, it is not always necessary to use official MLA format (though this should be taught at some point). Students may properly cite references in many ways, but these references must be in writing, with the original assignment. Therefore, the **answer is (A)**.

77. Animals with a notochord or a backbone are in the phylum

A. arthropoda.

B. chordata.

C. mollusca.

D. mammalia.

B. Chordata.

The phylum arthropoda contains spiders and insects and phylum mollusca contain snails and squid. Mammalia is a class in the phylum chordata. The **answer is (B).**

78. Which of the following is a correct explanation for scientific 'evolution'?

A. Giraffes need to reach higher for leaves to eat, so their necks stretch. The giraffe babies are then born with longer necks. Eventually, there are more long-necked giraffes in the population.

B. Giraffes with longer necks are able to reach more leaves, so they eat more and have more babies than other giraffes. Eventually, there are more long-necked giraffes in the population.

C. Giraffes want to reach higher for leaves to eat, so they release enzymes into their bloodstream, which in turn causes fetal development of longer-necked giraffes. Eventually, there are more long-necked giraffes in the population.

D. Giraffes with long necks are more attractive to other giraffes, so they get the best mating partners and have more babies. Eventually, there are more long-necked giraffes in the population.

B. Giraffes with longer necks are able to reach more leaves, so they eat more and have more babies than other giraffes. Eventually, there are more long-necked giraffes in the population.

Although evolution is often misunderstood, it occurs via natural selection. Organisms with a life/reproductive advantage will produce more offspring. Over many generations, this changes the proportions of the population. In any case, it is impossible for a stretched neck (A) or a fervent desire (C) to result in a biologically mutated baby. Although there are traits that are naturally selected because of mate attractiveness and fitness (D), this is not the primary situation here, **so answer (B) is the best choice**.

79. Which of the following is a correct definition for 'chemical equilibrium'?

A. Chemical equilibrium is when the forward and backward reaction rates are equal. The reaction may continue to proceed forward and backward.

B. Chemical equilibrium is when the forward and backward reaction rates are equal, and equal to zero. The reaction does not continue.

C. Chemical equilibrium is when there are equal quantities of reactants and products.

D. Chemical equilibrium is when acids and bases neutralize each other fully.

A. Chemical equilibrium is when the forward and backward reaction rates are equal. The reaction may continue to proceed forward and backward.

Chemical equilibrium is defined as when the quantities of reactants and products are at a 'steady state' and are no longer shifting, but the reaction may still proceed forward and backward. The rate of forward reaction must equal the rate of backward reaction. Note that there may or may not be equal amounts of chemicals, and that this is not restricted to a completed reaction or to an acid-base reaction. Therefore, the **answer is (A)**.

80. Which of the following data sets is properly represented by a bar graph?

A. Number of people choosing to buy cars, vs. Color of car bought.

B. Number of people choosing to buy cars, vs. Age of car customer.

C. Number of people choosing to buy cars, vs. Distance from car lot to customer home.

D. Number of people choosing to buy cars, vs. Time since last car purchase.

A. Number of people choosing to buy cars, vs. Color of car bought.

A bar graph should be used only for data sets in which the independent variable is non-continuous (discrete), e.g. gender, color, etc. Any continuous independent variable (age, distance, time, etc.) should yield a scatter-plot when the dependent variable is plotted. Therefore, the **answer must be (A)**.

81. In a science experiment, a student needs to dispense very small measured amounts of liquid into a well-mixed solution. Which of the following is the \best choice for his/her equipment to use?

A. Buret with Buret Stand, Stir-plate, Stirring Rod, Beaker.

B. Buret with Buret Stand, Stir-plate, Beaker.

C. Volumetric Flask, Dropper, Graduated Cylinder, Stirring Rod.

D. Beaker, Graduated Cylinder, Stir-plate.

B. Buret with Buret Stand, Stir-plate, Beaker.

The most accurate and convenient way to dispense small measured amounts of liquid in the laboratory is with a buret, on a buret stand. To keep a solution well-mixed, a magnetic stir-plate is the most sensible choice, and the solution will usually be mixed in a beaker. Although other combinations of materials could be used for this experiment, **choice (B)** is thus the simplest and best.

82. A laboratory balance is most appropriately used to measure the mass of which of the following?

A. Seven paper clips.

B. Three oranges.

C. Two hundred cells.

D. One student's elbow.

A. Seven paper clips.

Usually, laboratory/classroom balances can measure masses between approximately 0.01 gram and 1 kilogram. Therefore, answer (B) is too heavy and answer (C) is too light. Answer (D) is silly, but it is a reminder to instruct students not to lean on the balances or put their things near them. **Answer (A)**, which is likely to have a mass of a few grams, is correct in this case.

83. All of the following are measured in units of length, *except* for:

A. Perimeter.

B. Distance.

C. Radius.

D. Area.

D. Area.

Perimeter is the distance around a shape; distance is equivalent to length; radius is the distance from the center (e.g. in a circle) to the edge. Area, however, is the squared-length-units measure of the size of a two-dimensional surface. Therefore, **the answer is (D)**.

84. What is specific gravity?

A. The mass of an object.

B. The ratio of the density of a substance to the density of water.

C. Density.

D. The pull of the earth's gravity on an object.

B. The ratio of the density of a substance to the density of water.

Mass is a measure of the amount of matter in an object. Density is the mass of a substance contained per unit of volume. Weight is the measure of the earth's pull of gravity on an object. The only option here is the ratio of the density of a substance to the density of water, **answer (B).**

85. What is the most accurate description of the Water Cycle?

A. Rain comes from clouds, filling the ocean. The water then evaporates and becomes clouds again.

B. Water circulates from rivers into groundwater and back, while water vapor circulates in the atmosphere.

C. Water is conserved except for chemical or nuclear reactions, and any drop of water could circulate through clouds, rain, ground-water, and surface-water.

D. Weather systems cause chemical reactions to break water into its atoms.

C. **Water is conserved except for chemical or nuclear reactions, and any drop of water could circulate through clouds, rain, ground-water, and surface-water.**

All natural chemical cycles, including the Water Cycle, depend on the principle of Conservation of Mass. (For water, unlike for elements such as Nitrogen, chemical reactions may cause sources or sinks of water molecules.) Any drop of water may circulate through the hydrologic system, ending up in a cloud, as rain, or as surface- or ground-water. Although answers (A) and (B) describe parts of the water cycle, the most comprehensive and correct **answer is (C)**.

86. The scientific name *Canis familiaris* refers to the animal's

____

A. kingdom and phylum.

B. genus and species.

C. class and species.

D. type and family.

B. **Genus and species.**
To answer this question, you must be aware that genus and species are the most specific way to identify an organism, and that usually the genus is capitalized and the species, immediately following, is not. Furthermore, it helps to recall that 'Canis' is the genus for dogs, or canines. Therefore, the **answer must be (B)**. If you did not remember these details, you might recall that there is no such kingdom as 'Canis,' and that there isn't a category 'type' in official taxonomy. This could eliminate answers (A) and (D).

87. Members of the same animal species ________________

A. look identical.

B. never adapt differently.

C. are able to reproduce with one another.

D. are found in the same location.

C. Are able to reproduce with one another.

Although members of the same animal species may look alike (A), adapt alike (B), or be found near one another (D), the only requirement is that they be able to reproduce with one another. This ability to reproduce within the group is considered the hallmark of a species. Therefore, the **answer is (C)**.

88. Which of the following is/are true about scientists?

I. Scientists usually work alone.
II. Scientists usually work with other scientists.
III. Scientists achieve more prestige from new discoveries than from replicating established results.
IV. Scientists keep records of their own work, but do not publish it for outside review.

A. I and IV only.

B. II only.

C. II and III only.

D. III and IV only.

C. II and III only.

In the scientific community, scientists nearly always work in teams, both within their institutions and across several institutions. This eliminates (I) and requires (II), leaving only **answer choices (B) and (C)**. Scientists do achieve greater prestige from new discoveries, so the answer must be (C). Note that scientists must publish their work in peer-reviewed journals, eliminating (IV) in any case.

89. What is necessary for ion diffusion to occur spontaneously?

A. Carrier proteins.

B. Energy from an outside source.

C. A concentration gradient.

D. Cell flagellae.

C. A concentration gradient.

Spontaneous diffusion occurs when random motion leads particles to increase entropy by equalizing concentrations. Particles tend to move into places of lower concentration. Therefore, a concentration gradient is required, and the **answer is (C)**. No proteins (A), outside energy (B), or flagellae (D) are required for this process.

90. All of the following are considered Newton's Laws *except* for:

A. An object in motion will continue in motion unless acted upon by an outside force.

B. For every action force, there is an equal and opposite reaction force.

C. Nature abhors a vacuum.

D. Mass can be considered the ratio of force to acceleration.

C. Nature abhors a vacuum.

Newton's Laws include his law of inertia (an object in motion (or at rest) will stay in motion (or at rest) until acted upon by an outside force) (A), his law that (Force)=(Mass)(Acceleration) (D), and his equal and opposite reaction force law (B). Therefore, the **answer to this question is (C)**, because "Nature abhors a vacuum" is not one of these.

91. A cup of hot liquid and a cup of cold liquid are both sitting in a room at comfortable room temperature and humidity. Both cups are thin plastic. Which of the following is a true statement?

A. There will be fog on the outside of the hot liquid cup, and also fog on the outside of the cold liquid cup.

B. There will be fog on the outside of the hot liquid cup, but not on the cold liquid cup.

C. There will be fog on the outside of the cold liquid cup, but not on the hot liquid cup.

D. There will not be fog on the outside of either cup.

C. There will be fog on the outside of the cold liquid cup, but not on the hot liquid cup.

Fog forms on the outside of a cup when the contents of the cup are colder than the surrounding air, and the cup material is not a perfect insulator. This happens because the air surrounding the cup is cooled to a lower temperature than the ambient room, so it has a lower saturation point for water vapor. Although the humidity had been reasonable in the warmer air, when that air circulates near the colder region and cools, water condenses onto the cup's outside surface. This phenomenon is also visible when someone takes a hot shower, and the mirror gets foggy. The mirror surface is cooler than the ambient air, and provides a surface for water condensation. Furthermore, the same phenomenon is why defrosters on car windows send heat to the windows—the warmer window does not permit as much condensation. Therefore, the correct **answer is (C)**.

92. A ball rolls down a smooth hill. You may ignore air resistance. Which of the following is a true statement?

A. The ball has more energy at the start of its descent than just before it hits the bottom of the hill, because it is higher up at the beginning.

B. The ball has less energy at the start of its descent than just before it hits the bottom of the hill, because it is moving more quickly at the end.

C. The ball has the same energy throughout its descent, because positional energy is converted to energy of motion.

D. The ball has the same energy throughout its descent, because a single object (such as a ball) cannot gain or lose energy.

C. The ball has the same energy throughout its descent, because positional energy is converted to energy of motion.

The principle of Conservation of Energy states that (except in cases of nuclear reaction, when energy may be created or destroyed by conversion to mass), "Energy is neither created nor destroyed, but may be transformed." Answers (A) and (B) give you a hint in this question—it is true that the ball has more Potential Energy when it is higher, and that it has more Kinetic Energy when it is moving quickly at the bottom of its descent. However, the total sum of all kinds of energy in the ball remains constant, if we neglect 'losses' to heat/friction. Note that a single object can and does gain or lose energy when the energy is transferred to or from a different object. Conservation of Energy applies to systems, not to individual objects unless they are isolated. Therefore, the **answer must be (C)**.

93. A long silver bar has a temperature of 50 degrees Celsius at one end and 0 degrees Celsius at the other end. The bar will reach thermal equilibrium (barring outside influence) by the process of heat _________.

A. conduction.

B. convection.

C. radiation.

D. phase change.

A. conduction.

Heat conduction is the process of heat transfer via solid contact. The molecules in a warmer region vibrate more rapidly, jostling neighboring molecules and accelerating them. This is the dominant heat transfer process in a solid with no outside influences. Recall, also, that convection is heat transfer by way of fluid currents; radiation is heat transfer via electromagnetic waves; phase change can account for heat transfer in the form of shifts in matter phase. The answer to this question must **therefore be (A)**.

94. ___________ are cracks in the plates of the earth's crust, along which the plates move.

A. Faults

B. Ridges

C. Earthquakes

D. Volcanoes

A. Faults.

Faults are cracks in the earth's crust, and when the earth moves, an earthquake results. Faults may lead to mismatched edges of ground, forming ridges, and ground shape may also be determined by volcanoes. The answer to this question must **therefore be (A).**

95. Fossils are usually found in ____________ rock.

A. igneous.

B. sedimentary.

C. metamorphic.

D. cumulus.

B. Sedimentary

Fossils are formed by layers of dirt and sand settling around organisms, hardening, and taking an imprint of the organisms. When the organism decays, the hardened imprint is left behind. This is most likely to happen in rocks that form from layers of settling dirt and sand, i.e. sedimentary rock. Note that igneous rock is formed from molten rock from volcanoes (lava), while metamorphic rock can be formed from any rock under very high temperature and pressure changes. 'Cumulus' is a descriptor for clouds, not rocks. The best answer is **therefore (B)**.

96. Which of the following is *not* a common type of acid in 'acid rain' or acidified surface water?

A. Nitric acid.

B. Sulfuric acid.

C. Carbonic acid.

D. Hydrofluoric acid.

D. **Hydrofluoric acid.**

Acid rain forms predominantly from pollutant oxides in the air (usually nitrogen-based NO_x or sulfur-based SO_x), which become hydrated into their acids (nitric or sulfuric acid). Because of increased levels of carbon dioxide pollution, carbonic acid is also common in acidified surface water environments. Hydrofluoric acid can be found, but it is much less common. In general, carbon, nitrogen, and sulfur are much more prevalent in the environment than fluorine. Therefore, the **answer is (D)**.

97. Which of the following is *not* true about phase change in matter?

A. Solid water and liquid ice can coexist at water's freezing point.

B. At 7 degrees Celsius, water is always in liquid phase.

C. Matter changes phase when enough energy is gained or lost.

D. Different phases of matter are characterized by differences in molecular motion.

B. At 7 degrees Celsius, water is always in liquid phase.

According to the molecular theory of matter, molecular motion determines the 'phase' of the matter, and the energy in the matter determines the speed of molecular motion. Solids have vibrating molecules that are in fixed relative positions; liquids have faster molecular motion than their solid forms, and the molecules may move more freely but must still be in contact with one another; gases have even more energy and more molecular motion. (Other phases, such as plasma, are yet more energetic.) At the 'freezing point' or 'boiling point' of a substance, both relevant phases may be present. For instance, water at zero degrees Celsius may be composed of some liquid and some solid, or all liquid, or all solid. Pressure changes, in addition to temperature changes, can cause phase changes. For example, nitrogen can be liquefied under high pressure, even though its boiling temperature is very low. Therefore, the **correct answer must be (B)**. Water may be a liquid at that temperature, but it may also be a solid, depending on ambient pressure.

98. Which of the following is the longest (largest) unit of geological time?

A. Solar Year.

B. Epoch.

C. Period.

D. Era.

D. **Era**.

Geological time is measured by many units, but the longest unit listed here (and indeed the longest used to describe the biological development of the planet) is the Era. Eras are subdivided into Periods, which are further divided into Epochs. Therefore, the **answer is (D).**

99. Extensive use of antibacterial soap has been found to increase the virulence of certain infections in hospitals. Which of the following might be an explanation for this phenomenon?

A. Antibacterial soaps do not kill viruses.

B. Antibacterial soaps do not incorporate the same antibiotics used as medicine.

C. Antibacterial soaps kill a lot of bacteria, and only the hardiest ones survive to reproduce.

D. Antibacterial soaps can be very drying to the skin.

C. **Antibacterial soaps kill a lot of bacteria, and only the hardiest ones survive to reproduce**.

All of the answer choices in this question are true statements, but the question specifically asks for a cause of increased disease virulence in hospitals. This phenomenon is due to natural selection. The bacteria that can survive contact with antibacterial soap are the strongest ones, and without other bacteria competing for resources, they have more opportunity to flourish. This problem has led to several antibiotic-resistant bacterial diseases in hospitals nation-wide. Therefore, the **answer is (C)**. However, note that answers (A) and (D) may be additional problems with over-reliance on antibacterial products.

100. Which of the following is a correct explanation for astronaut 'weightlessness'?

A. Astronauts continue to feel the pull of gravity in space, but they are so far from planets that the force is small.

B. Astronauts continue to feel the pull of gravity in space, but spacecraft have such powerful engines that those forces dominate, reducing effective weight.

C. Astronauts do not feel the pull of gravity in space, because space is a vacuum.

D. Astronauts do not feel the pull of gravity in space, because black hole forces dominate the force field, reducing their masses.

A. Astronauts continue to feel the pull of gravity in space, but they are so far from planets that the force is small.

Gravity acts over tremendous distances in space (theoretically, infinite distance, though certainly at least as far as any astronaut has traveled). However, gravitational force is inversely proportional to distance squared from a massive body. This means that when an astronaut is in space, s/he is far enough from the center of mass of any planet that the gravitational force is very small, and s/he feels 'weightless'. Space is mostly empty (i.e. vacuum), and there are some black holes, and spacecraft do have powerful engines. However, none of these has the effect attributed to it in the incorrect answer choices (B), (C), or (D). The answer to this question must **therefore be (A).**

101. The theory of 'sea floor spreading' explains _______________

A. the shapes of the continents.

B. how continents collide.

C. how continents move apart.

D. how continents sink to become part of the ocean floor.

C. How continents move apart.

In the theory of 'sea floor spreading', the movement of the ocean floor causes continents to spread apart from one another. This occurs because crust plates split apart, and new material is added to the plate edges. This process pulls the continents apart, or may create new separations, and is believed to have caused the formation of the Atlantic Ocean. The **answer is (C)**.

102. Which of the following animals are most likely to live in a tropical rain forest?

A. Reindeer.

B. Monkeys.

C. Puffins.

D. Bears.

B. Monkeys.

The tropical rain forest biome is hot and humid, and is very fertile—it is thought to contain almost half of the world's species. Reindeer (A), puffins (C), and bears (D), however, are usually found in much colder climates. There are several species of monkeys that thrive in hot, humid climates, so **answer (B) is correct.**

103. Which of the following is *not* a type of volcano?

A. Shield Volcanoes.

B. Composite Volcanoes.

C. Stratus Volcanoes.

D. Cinder Cone Volcanoes.

C. Stratus Volcanoes.

There are three types of volcanoes. Shield volcanoes (A) are associated with non-violent eruptions and repeated lava flow over time. Composite volcanoes (B) are built from both lava flow and layers of ash and cinders. Cinder cone volcanoes (D) are associated with violent eruptions, such that lava is thrown into the air and becomes ash or cinder before falling and accumulating. **'Stratus' (C)** is a type of cloud, not volcano, so it is the correct answer to this question.

104. Which of the following is *not* a property of metalloids?

A. Metalloids are solids at standard temperature and pressure.

B. Metalloids can conduct electricity to a limited extent.

C. Metalloids are found in groups 13 through 17.

D. Metalloids all favor ionic bonding.

D. **Metalloids all favor ionic bonding**.

Metalloids are substances that have characteristics of both metals and nonmetals, including limited conduction of electricity and solid phase at standard temperature and pressure. Metalloids are found in a 'stair-step' pattern from Boron in group 13 through Astatine in group 17. Some metalloids, e.g. Silicon, favor covalent bonding. Others, e.g. Astatine, can bond ionically. Therefore, **the answer is (D).** Recall that metals/nonmetals/metalloids are not strictly defined by Periodic Table group, so their bonding is unlikely to be consistent with one another.

105. Which of these is a true statement about loamy soil?

A. Loamy soil is gritty and porous.

B. Loamy soil is smooth and a good barrier to water.

C. Loamy soil is hostile to microorganisms.

D. Loamy soil is velvety and clumpy.

D. **Loamy soil is velvety and clumpy**.

The three classes of soil by texture are: Sandy (gritty and porous), Clay (smooth, greasy, and most impervious to water), and Loamy (velvety, clumpy, and able to hold water and let water flow through). In addition, loamy soils are often the most fertile soils. Therefore, the **answer must be (D)**.

106. Lithification refers to the process by which unconsolidated sediments are transformed into ______________

A. metamorphic rocks.

B. sedimentary rocks.

C. igneous rocks.

D. lithium oxide.

B. **Sedimentary rocks.**

Lithification is the process of sediments coming together to form rocks, i.e. sedimentary rock formation. Metamorphic and igneous rocks are formed via other processes (heat and pressure or volcano, respectively). Lithium oxide shares a word root with 'lithification' but is otherwise unrelated to this question. Therefore, the **answer must be (B)**.

107. Igneous rocks can be classified according to which of the following?

A. Texture.

B. Composition.

C. Formation process.

D. All of the above.

D. **All of the above**.

Igneous rocks, which form from the crystallization of molten lava, are classified according to many of their characteristics, including texture, composition, and how they were formed. Therefore, **the answer is (D).**

108. Which of the following is the most accurate definition of a nonrenewable resource?

A. A nonrenewable resource is never replaced once used.

B. A nonrenewable resource is replaced on a timescale that is very long relative to human life-spans.

C. A nonrenewable resource is a resource that can only be manufactured by humans.

D. A nonrenewable resource is a species that has already become extinct.

B. A nonrenewable resource is replaced on a timescale that is very long relative to human life-spans.

Renewable resources are those that are renewed, or replaced, in time for humans to use more of them. Examples include fast-growing plants, animals, or oxygen gas. (Note that while sunlight is often considered a renewable resource, it is actually a nonrenewable but extremely abundant resource.) Nonrenewable resources are those that renew themselves only on very long timescales, usually geologic timescales. Examples include minerals, metals, or fossil fuels. Therefore, the **correct answer is (B)**.

109. The theory of 'continental drift' is supported by which of the following?

A. The way the shapes of South America and Europe fit together.

B. The way the shapes of Europe and Asia fit together.

C. The way the shapes of South America and Africa fit together.

D. The way the shapes of North America and Antarctica fit together.

C. The way the shapes of South America and Africa fit together.

The theory of 'continental drift' states that many years ago, there was one land mass on the earth ('pangea'). This land mass broke apart via earth crust motion, and the continents drifted apart as separate pieces. This is supported by the shapes of South America and Africa, which seem to fit together like puzzle pieces if you look at a globe. Note that answer choices (A), (B), and (D) give either land masses that do not fit together, or those that are still attached to each other. Therefore, the **answer must be (C)**.

110. When water falls to a cave floor and evaporates, it may deposit calcium carbonate. This process leads to the formation of which of the following?

A. Stalactites.

B. Stalagmites.

C. Fault lines.

D. Sedimentary rocks.

B. **Stalagmites**.

To answer this question, recall the trick to remember the kinds of crystals formed in caves. Stalactites have a 'T' in them, because they form hanging from the ceiling (resembling a 'T'). Stalagmites have an 'M' in them, because they make bumps on the floor (resembling an 'M'). Note that fault lines and sedimentary rocks are irrelevant to this question. Therefore, **the answer must be (B)**.

111. A child has type O blood. Her father has type A blood, and her mother has type B blood. What are the genotypes of the father and mother, respectively?

A. AO and BO.

B. AA and AB.

C. OO and BO.

D. AO and BB.

A. AO and BO.

Because O blood is recessive, the child must have inherited two O's—one from each of her parents. Since her father has type A blood, his genotype must be AO; likewise her mother's blood must be BO. Therefore, only **answer (A)** can be correct.

112. Which of the following is the best definition for 'meteorite'?

A. A meteorite is a mineral composed of mica and feldspar.

B. A meteorite is material from outer space, that has struck the earth's surface.

C. A meteorite is an element that has properties of both metals and nonmetals.

D. A meteorite is a very small unit of length measurement.

B. A meteorite is material from outer space, that has struck the earth's surface.

Meteoroids are pieces of matter in space, composed of particles of rock and metal. If a meteoroid travels through the earth's atmosphere, friction causes burning and a "shooting star"—i.e. a meteor. If the meteor strikes the earth's surface, it is known as a meteorite. Note that although the suffix –ite often means a mineral, answer (A) is incorrect. Answer (C) refers to a 'metalloid' rather than a 'meteorite', and answer (D) is simply a misleading pun on 'meter'. Therefore, the **answer is (B)**.

113. A white flower is crossed with a red flower. Which of the following is a sign of incomplete dominance?

A. Pink flowers.

B. Red flowers.

C. White flowers.

D. No flowers.

A. Pink flowers.

Incomplete dominance means that neither the red nor the white gene is strong enough to suppress the other. Therefore both are expressed, leading in this case to the formation of pink flowers. Therefore, the **answer is (A)**.

114. What is the source for most of the United States' drinking water?

A. Desalinated ocean water.

B. Surface water (lakes, streams, mountain runoff).

C. Rainfall into municipal reservoirs.

D. Groundwater.

D. Groundwater.

Groundwater currently provides drinking water for 53% of the population of the United States. (Although groundwater is often less polluted than surface water, it can be contaminated and it is very hard to clean once it is polluted. If too much groundwater is used from one area, then the ground may sink or shift, or local salt water may intrude from ocean boundaries.) The other answer choices can be used for drinking water, but they are not the most widely used. Therefore, **the answer is (D)**.

115. Which is the correct sequence of insect development?

A. Egg, pupa, larva, adult.

B. Egg, larva, pupa, adult.

C. Egg, adult, larva, pupa.

D. Pupa, egg, larva, adult.

B. Egg, larva, pupa, adult.

An insect begins as an egg, hatches into a larva (e.g. caterpillar), forms a pupa (e.g. cocoon), and emerges as an adult (e.g. moth). Therefore, the **answer is (B)**.

116. A wrasse (fish) cleans the teeth of other fish by eating away plaque. This is an example of ____________________ between the fish.

A. parasitism.

B. symbiosis (mutualism).

C. competition.

D. predation.

B. Symbiosis (mutualism).

When both species benefit from their interaction in their habitat, this is called 'symbiosis', or 'mutualism'. In this example, the wrasse benefits from having a source of food, and the other fish benefit by having healthier teeth. Note that 'parasitism' is when one species benefits at the expense of the other, 'competition' is when two species compete with one another for the same habitat or food, and 'predation' is when one species feeds on another. Therefore, the **answer is (B)**.

117. What is the main obstacle to using nuclear fusion for obtaining electricity?

A. Nuclear fusion produces much more pollution than nuclear fission.

B. There is no obstacle; most power plants us nuclear fusion today.

C. Nuclear fusion requires very high temperature and activation energy.

D. The fuel for nuclear fusion is extremely expensive.

C. Nuclear fusion requires very high temperature and activation energy.

Nuclear fission is the usual process for power generation in nuclear power plants. This is carried out by splitting nuclei to release energy. The sun's energy is generated by nuclear fusion, i.e. combination of smaller nuclei into a larger nucleus. Fusion creates much less radioactive waste, but it requires extremely high temperature and activation energy, so it is not yet feasible for electricity generation. Therefore, the **answer is (C)**.

118. Which of the following is a true statement about radiation exposure and air travel?

A. Air travel exposes humans to radiation, but the level is not significant for most people.

B. Air travel exposes humans to so much radiation that it is recommended as a method of cancer treatment.

C. Air travel does not expose humans to radiation.

D. Air travel may or may not expose humans to radiation, but it has not yet been determined.

A. Air travel exposes humans to radiation, but the level is not significant for most people.

Humans are exposed to background radiation from the ground and in the atmosphere, but these levels are not considered hazardous under most circumstances, and these levels have been studied extensively. Air travel does create more exposure to atmospheric radiation, though this is much less than people usually experience through dental X-rays or other medical treatment. People whose jobs or lifestyles include a great deal of air flight may be at increased risk for certain cancers from excessive radiation exposure. Therefore, the **answer is (A)**.

119. Which process(es) result(s) in a haploid chromosome number?

A. Mitosis.

B. Meiosis.

C. Both mitosis and meiosis.

D. Neither mitosis nor meiosis.

B. Meiosis.

Meiosis is the division of sex cells. The resulting chromosome number is half the number of parent cells, i.e. a 'haploid chromosome number'. Mitosis, however, is the division of other cells, in which the chromosome number is the same as the parent cell chromosome number. Therefore, the **answer is (B)**.

120. Which of the following is *not* a member of Kingdom Fungi?

A. Mold.

B. Blue-green algae.

C. Mildew.

D. Mushrooms.

B. Blue-green Algae.

Mold (A), mildew (C), and mushrooms (D) are all types of fungus. Blue-green algae, however, is in Kingdom Monera. Therefore, the **answer is (B)**.

121. Which of the following organisms use spores to reproduce?

A. Fish.

B. Flowering plants.

C. Conifers.

D. Ferns.

D. **Ferns.**

Ferns, in Division Pterophyta, reproduce with spores and flagellated sperm. Flowering plants reproduce via seeds, and conifers reproduce via seeds protected in cones (e.g. pinecone). Fish, of course, reproduce sexually. Therefore, the **answer is (D)**.

122. What is the main difference between the 'condensation hypothesis' and the 'tidal hypothesis' for the origin of the solar system?

A. The tidal hypothesis can be tested, but the condensation hypothesis cannot.

B. The tidal hypothesis proposes a near collision of two stars pulling on each other, but the condensation hypothesis proposes condensation of rotating clouds of dust and gas.

C. The tidal hypothesis explains how tides began on planets such as Earth, but the condensation hypothesis explains how water vapor became liquid on Earth.

D. The tidal hypothesis is based on Aristotelian physics, but the condensation hypothesis is based on Newtonian mechanics.

B. The tidal hypothesis proposes a near collision of two stars pulling on each other, but the condensation hypothesis proposes condensation of rotating clouds of dust and gas.

Most scientists believe the 'condensation hypothesis,' i.e. that the solar system began when rotating clouds of dust and gas condensed into the sun and planets. A minority opinion is the 'tidal hypothesis,' i.e. that the sun almost collided with a large star. The large star's gravitational field would have then pulled gases out of the sun; these gases are thought to have begun to orbit the sun and condense into planets. Because both of these hypotheses deal with ancient, unrepeatable events, neither can be tested, eliminating answer (A). Note that both 'tidal' and 'condensation' have additional meanings in physics, but those are not relevant here, eliminating answer (C). Both hypotheses are based on best guesses using modern physics, eliminating answer (D). Therefore, the **answer is (B)**.

123. Which of the following units is *not* a measure of distance?

A. AU (astronomical unit).

B. Light year.

C. Parsec.

D. Lunar year.

D. Lunar year.

Although the terminology is sometimes confusing, it is important to remember that a 'light year' (B) refers to the distance that light can travel in a year. Astronomical Units (AU) (A) also measure distance, and one AU is the distance between the sun and the earth. Parsecs (C) also measure distance, and are used in astronomical measurement- they are very large, and are usually used to measure interstellar distances. A lunar year, or any other kind of year for a planet or moon, is the *time* measure of that body's orbit. Therefore, the answer to this **question is (D)**.

124. The salinity of ocean water is closest to ____________ .

A. 0.035 %

B. 0.35 %

C. 3.5 %

D. 35 %

C. 3.5 %

Salinity, or concentration of dissolved salt, can be measured in mass ratio (i.e. mass of salt divided by mass of sea water). For Earth's oceans, the salinity is approximately 3.5 %, or 35 parts per thousand. Note that answers (A) and (D) can be eliminated, because (A) is so dilute as to be hardly saline, while (D) is so concentrated that it would not support ocean life. Therefore, the **answer is (C)**.

125. Which of the following will not change in a chemical reaction?

A. Number of moles of products.

B. Atomic number of one of the reactants.

C. Mass (in grams) of one of the reactants.

D. Rate of reaction.

B. Atomic number of one of the reactants.

Atomic number, i.e. the number of protons in a given element, is constant unless involved in a nuclear reaction. Meanwhile, the amounts (measured in moles (A) or in grams(C)) of reactants and products change over the course of a chemical reaction, and the rate of a chemical reaction (D) may change due to internal or external processes. Therefore, the **answer is (B)**.

XAMonline, INC. 21 Orient Ave. Melrose, MA 02176

Toll Free number 800-509-4128

TO ORDER Fax 781-662-9268 OR www.XAMonline.com

MASSACHUSETTS TEST FOR EDUCATOR LICENTURE - MTEL - 2007

P0# Store/School:

Address 1:

Address 2 (Ship to other):

City, State Zip

Credit card number______-_______-________-_______ expiration______

EMAIL ____________________________

PHONE FAX

13# ISBN 2007	TITLE	Qty	Retail	Total
978-1-58197-884-1	MTEL Biology 13			
978-1-58197-883-4	MTEL Chemistry 12			
978-1-58197-875-9	MTEL Communication and Literacy Skills 01			
978-1-58197-885-8	MTEL Earth Science 14			
978-1-58197-879-7	MTEL English 07			
978-1-58197-892-6	MTEL Foundations of Reading 90 (requirement all El. Ed)			
978-1-58197-887-2	MTEL French 26			
978-1-58197-876-6	MTEL General Curriculum (formerly Elementary) 03			
978-1-58197-877-3	MTEL General Curriculum (formerly Elementary) 03 Sample Questions			
978-1-58197-881-0	MTEL General Science 10			
978-1-58197-878-0	MTEL History 06 (Social Science)			
978-1-58197-196-5	MTEL Latin & Classical Humanities 15			
978-1-58197-880-3	MTEL Mathematics 09			
978-1-58197-890-2	MTEL Middle School Humanities 50			
978-1-58197-889-6	MTEL Middle School Mathematics 47			
978-1-58197-891-9	MTEL Middle School Mathematics-Science 51			
978-1-58197-886-5	MTEL Physical Education 22			
978-1-58197-882-7	MTEL Physics Sample Test 11			
978-1-58197-898-8	MTEL Political Science/Political Philosophy 48			
978-1-58197-888-9	MTEL Spanish 28			
			SUBTOTAL	
	FOR PRODUCT PRICES VISIT WWW.XAMONLINE.COM		Ship	$8.25
			TOTAL	

www.ingramcontent.com/pod-product-compliance
Lightning Source LLC
LaVergne TN
LVHW081642180625
814175LV00012B/429
* 9 7 8 1 5 8 1 9 7 8 9 1 9 *